Nanoscale Processes
on
Insulating Surfaces

Nanoscale Processes
on
Insulating Surfaces

Enrico Gnecco
University of Basel, Switzerland

Marek Szymonski
Jagiellonian University, Poland

World Scientific

NEW JERSEY · LONDON · SINGAPORE · BEIJING · SHANGHAI · HONG KONG · TAIPEI · CHENNAI

Published by

World Scientific Publishing Co. Pte. Ltd.

5 Toh Tuck Link, Singapore 596224

USA office: 27 Warren Street, Suite 401-402, Hackensack, NJ 07601

UK office: 57 Shelton Street, Covent Garden, London WC2H 9HE

British Library Cataloguing-in-Publication Data

A catalogue record for this book is available from the British Library.

ISBN-13 978-981-283-762-2

ISBN-10 981-283-762-0

About the authors

Dr. Enrico Gnecco was awarded his Ph.D. from the University of Genova in 2000. He started his research activities as an undergraduate student with investigations of current noise in Hg and Tl based high-temperature superconductors. His Ph.D. thesis focused on nanotribology, i.e. the study of friction and wear processes occurring on the nanometer scale. Within the group of Prof. Ugo Valbusa he characterized the growth of amorphous carbon films by scanning probe microscopy, and studied the relationship between friction and self-affine properties of these films. In the second part of his Ph.D. studies, he joined the group of Prof. Ernst Meyer in Basel, where he investigated friction processes on alkali halide surfaces in ultra high vacuum (UHV). The main result was the observation of a logarithmic velocity dependence of atomic friction, which was interpreted within a combination of the classical Tomlinson and Eyring models.

After his Ph.D. he joined the Swiss National Center of Competence of Research in Nanoscale Science, where he studied other basic nanotribological phenomena, such as abrasion wear on ionic crystals in UHV, various forms of superlubricity onset transitions, and electromechanical methods to reduce friction on the atomic scale. Most of these experiments were performed on insulating surfaces, and have found promising applications in the development of micro- and nano-devices. He also investigated self-assembly of organic molecules on insulators, using an original technique to trap molecules in rectangular nanopits produced by electron irradiation. He is also interested in surface nanomanipulation on different systems (from metal clusters to single molecules) with the goal of evaluating the dissipative forces accompanying the manipulation process.

E.G. has co-authored more than 40 scientific publications, including articles on Science, Nanoletters, Nanotechnology, and Physical Review. He was also awarded a diploma in piano from the Conservatory of Music of his hometown, Genova.

Prof. Marek Szymonski was awarded his Ph.D. from the Jagiellonian University, Krakow, in 1978. He joined the Physics Department of the Jagiellonian University in 1973, he was awarded a higher doctoral degree (so called 'habilitation') in 1982, and he was appointed to the full professor position at this University in 1991. He held several positions during his academic carrier, such as Vice-Rector of the Jagiellonian University (1993-1999), Chairman of the Supervisory Board for the 'Krakow Center for

Advanced Technologies' (1998-1999), Chairman, Scientific Council of the Regional Laboratory for Physicochemical Analysis, Jagiellonian University (1999-2006), Dean of the Faculty of Physics of the Jagiellonian University (2003-2005), and President of the Polish Vacuum Society (1996-1998 and 2004-2007). He also served several international posts, such as Councilor (1992-1998 and 2004-2007) and Alternate Councilor (2001-2004) to the Executive Council of the Int. Union for Vacuum Science, Technology and Applications and Chairman of the European Representative Committee of IUVSTA (2004-2007). Currently he is heading the Division of Physics of Nanostructures and Nanotechnology of the Institute of Physics, Jagiellonian University, and he is Director of the Research Center on Nanometer-Scale Science and Advanced Materials, NANOSAM, at the Jagiellonian University.

His scientific record includes over 130 scientific publications and several dozens of invited lectures at scientific conferences. He was Ph.D. thesis supervisor for 12 Ph.D. students and for many M.Sc. graduates. He organized several national and international research workshops and conferences, among them the 20th European Conference on Surface Science (ECOSS-20), 2001, and the 12th International Conference on Organized Molecular Films (LB 12), 2007.

His research activities are concentrated on those topics of materials research which are related to properties of nano-size materials with particular focus on scanning probe methods, to design and characterization of functional materials with emphasis on semiconductors, dielectrics and molecular structures, to quantum phenomena in mesoscopic systems, to manufacturing and characterization of self-assembling structures, and to research on biomedical materials. His interest in ionic insulators dates back to 1977 when he participated in the pioneering experiments on electron-stimulated desorption of alkali halides together with Hans Overeijnder and Dolf de Vries from the FOM Institute of Atomic and Molecular Physics in Amsterdam.

Preface

The rapid development of scanning probe microscopy (SPM) has made possible investigations of morphological and physical properties of insulating surfaces with unprecedented resolution. Since the 'father' of the SPM family — the scanning tunneling microscope (STM) — can only be applied to image ultrathin insulating films grown on conducting surfaces, most of the SPM investigations on insulators have been performed using the atomic force microscope (AFM). The principle of AFM is simple: an ultrasharp tip is driven over a surface, and the tip-surface interaction is reconstructed by monitoring the deflection of a flexible cantilever supporting the tip. Thanks to the extraordinary sensitivity of the piezo-elements used for positioning the tip with respect to the surface, atomically resolved images of several insulating crystal surfaces can be readily obtained in such a way.

Before introducing the AFM and STM techniques, we briefly discuss the crystallographic structures and preparation methods of various insulating surfaces in the first two chapters. Our attention will be limited to alkali halide surfaces and to oxide surfaces with large band gaps. We will not address other insulating surfaces, such as those of polymers, plastics, glasses and minerals like mica, where the interpretation of the SPM results is often not so conclusive. We will also not cover issues like chemical reactivity of the surfaces except when they become relevant to interpret SPM images.

Alkali halide surfaces have a simple structure and can be easily prepared and characterized in an ultra high vacuum environment (UHV). For these reasons, they have quickly become reference models to investigate crystal growth processes on the nanometer scale. Alkali halide surfaces are also important playgrounds for nanoscale phenomena such as self-assembly of metal clusters and large organic molecules, or friction and wear processes.

Insulating oxide surfaces find a broad use in several applications, ranging from interfaces for electronic ceramics to chemical catalysis. Even if they are more difficult to characterize, these surfaces are still amenable to fundamental investigations on the nanoscale, and high-resolution SPM images can be obtained after the surfaces have been carefully prepared.

The central part of the book begins with Chapters 4 and 5. Here we will focus on bulk and ultrathin insulating surfaces as imaged by SPM. Atomic resolution is unquestionable when single defects are imaged, which is now commonly achieved on several structures studied by AFM. Some defects like vacancies can be even created and subsequently imaged by the same AFM tip, which gives important information on the scanning process itself. Several experimental results have been also complemented by theoretical simulations of the imaging process, which allowed to identify the main forces responsible for atomic resolution.

Chapters 6, 7 and 8 introduce the interaction of ions, electrons and photons with halide surfaces, with special emphasis on the nanostructures created by the interaction processes. The discussion of the basic mechanisms responsible for crystal erosion and large surface nanopatterning with nanometer precision is supplemented by recent experimental results obtained by means of high resolution AFM imaging.

Chapters 9 and 10 deal with self-assembly of metals and organic molecules on bare and nanopatterned insulating surfaces. Once again the discussion of the experimental results is complemented by theoretical interpretations of the imaging process. Since the ordering of metal and molecular adsorbates is often hindered by the weak interaction between adsorbate and substrate, nanopatterns play an important role in improving the stability of the adsorbed species. For instance, self-assembly can be readily achieved along monatomic step edges or inside nanometer-sized pits produced by electron irradiation. As a further step, connecting well-defined molecular assemblies to external electrodes via metal nanowires grown on insulating surfaces might become feasible in the near future.

In Chapter 11 we discuss force spectroscopy measurements on insulating surfaces. In such cases, the response of the SPM tip is monitored at different separations between tip and surface, which gives important information on the tip-surface interaction. If a current flow between tip and sample can be established, by decreasing the band gap in the material, scanning tunneling spectroscopy (STS) is also possible. With this technique metals and organic molecules deposited on thin insulating films can be also investigated, and different electronic states of single molecules have been even identified.

The last two chapters deal with mechanical phenomena induced and observed using SPM on insulating surfaces. These processes include friction, wear, indentation and manipulation of tiny nano-objects. While several important results have been obtained in the first three topics — last but not least the achievement of superlubricity — nanomanipulation on insulating surfaces is still in its embryonic phase. However, nanomanipulation has such potential applications to molecular electronics and nanomechanics that exciting experiments are on sight, once again driven by theoretical models.

This book could have not been written without the collaboration of several people who supported us during the time of writing. Even if it is not possible to cite all of them, a special thanks goes to Jacek J. Kolodziej, Franciszek Krok, Bartosz Such, Piotr Cyganik, Piotr Goryl, Piotr Piatkowski, Janusz Budzioch and Salah Raza Saeed from the Jagiellonian University of Krakow, and Ernst Meyer, Roland Bennewitz, Luca Ramoino, Sabine Maier, Laurent Nony, Anisoara Socoliuc, Alexis Baratoff, Thilo Glatzel, Lars Zimmerli, Oliver Pfeiffer, Akshata Rao, Pascal Steiner and Raphael Roth from the University of Basel. We also thank Urszula Lustofin for administrative assistance and Tim Ashworth for carefully reviewing parts of the book. E.G. would like to thank his wife Tatiana for her patience and for continuous encouragement while writing this book. Financial support from the European Community under the Maria Curie Host Fellowship for Transfer of Knowledge (contract no. MTKD-CT-2004-003132), the Polish Foundation for Science (contract for subsidy no. 11/2007), and the Swiss National Science Foundation and NCCR Nanoscale Science is also gratefully acknowledged.

E. Gnecco and M. Szymonski

Contents

Chapter 1

Crystal Structures of Insulating Surfaces

This chapter introduces the crystal structures and the main properties of several insulating halide and oxide surfaces that have been addressed by scanning probe microscopy on the nanometer scale. In the first part of the chapter we distinguish between alkali and alkaline earth halides. In the second part we introduce oxides surfaces, and divide them into true insulators and mixed conductors. The preparation methods of these surfaces are discussed in Chapter 2.

1.1 Halide Surfaces

1.1.1 *Alkali halide surfaces*

Alkali halides result from the binding of alkali metal to halogen ions. In standard room conditions alkali halides are white or transparent crystals. The most representative among them is *sodium chloride* (NaCl). Apart from being a material of obvious importance in several aspects of everyday life, NaCl plays a vital role in chemistry, biology, and several other scientific disciplines. Sodium chloride crystallizes in the *rock salt* structure shown in Fig. 1.1(a), which is common to all alkali halide crystals with the exceptions of CsCl, CsBr, and CsI [Ashcroft and Mermin (1976)]. Only very few facets of the NaCl structure are stable, in particular the (001) surface. This is due to the arrangement of the Na^+ and Cl^- ions, which makes such a facet electrically neutral. It is also well established that the ions at the surface undergo only small relaxations with respect to their bulk-truncated positions [Tasker (1979)].

In general, stable alkali halide surfaces are obtained by *cleavage*, i.e., by splitting a crystal along a definite plane. This process is always

1

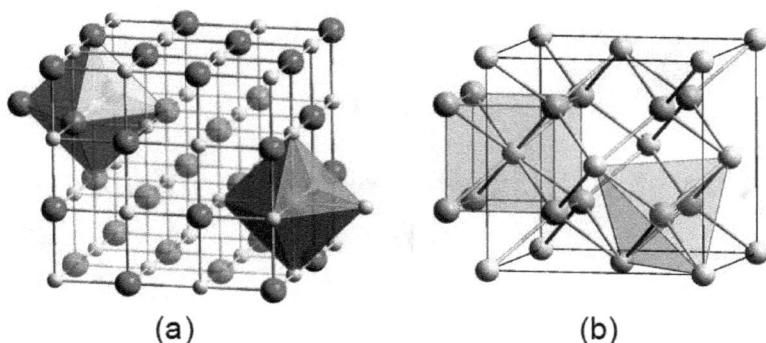

Fig. 1.1 Crystal structures of (a) sodium chloride and (b) calcium fluoride. Brighter spheres in (b) represent Ca^{2+} and darker spheres correspond to F^- ions.

accompanied by the formation of *steps*, which play an important role in many physical and chemical processes such as self-assembly of metal clusters and of large organic molecules. Characteristic nanopatterns can also be obtained by photon or electron irradiation or by ion bombardment of the surfaces, as discussed in details in Chapters 7 and 8.

1.1.2 *Alkaline earth halide surfaces*

Alkaline earth halides are other ionic materials of great interest in nanoscience. These crystals are formed by alkaline earth metal and halogen ions. The most studied among alkaline earth halides is *calcium fluoride* (CaF_2). Calcium fluoride crystallizes in the *fluorite* structure, which is shown in Fig. 1.1(b). In the fluorite structure each F^- ion is surrounded by four Ca^{2+} ions. Although the pure material is colorless, the mineral is often deeply colored due to the presence of *F-centers*, i.e. crystallographic defects in which a halogen vacancy in the crystal is filled by an electron. The most stable facet of the CaF_2 crystal is the (111) surface. This surface has a small lattice misfit of 0.6% with respect to Si(111), which makes calcium fluoride quite attractive as epitaxial insulator (Sec. 5.2).

1.2 Oxide Surfaces

Oxides are chemical compounds of oxygen with more electropositive elements or groups. Oxides have heterogeneous and complicated surfaces, which makes their fundamental properties quite difficult to recognize. In the following we will distinguish between *true insulating oxides* and *mixed conducting oxides* [Fu and Wagner (2007)]. While true insulating oxides are characterized by very large band gaps making their electric conductivity practically negligible, mobile electronic and ionic defects can be generated in mixed conducting oxides, according to distinct reactions, so that these solids can exhibit a certain electric and ionic conductivity.

1.2.1 *True insulating oxide surfaces*

1.2.1.1 *Aluminum oxide*

Aluminum oxide or *alumina* (Al_2O_3) has a band gap of 8.8 eV. Aluminum oxide has a narrow range of stoichiometry, and bulk defects do not increase its electronic conductivity. For these reasons Al_2O_3 is widely used as a catalyst support and as a substrate for growth of metal films. The most common way by which aluminum oxide crystallizes is the *corundum* structure shown in Fig. 1.2, which is also known as *α-aluminum oxide*. In this structure each unit cell contains six formula units of Al_2O_3 and the oxygen atoms nearly form a hexagonal close-packed substructure with aluminum atoms filling two thirds of octahedral interstices. *Rubies* and *sapphires* are gem-quality forms of corundum with their characteristic colors due to impurities in the crystal structure. The most stable among the unreconstructed alumina surfaces is the $α$-$Al_2O_3(0001)$ surface terminated by a single Al layer. This surface undergoes a series of reconstructions upon annealing at high temperatures in vacuum and oxygen desorption [French and Somorjai (1970)].

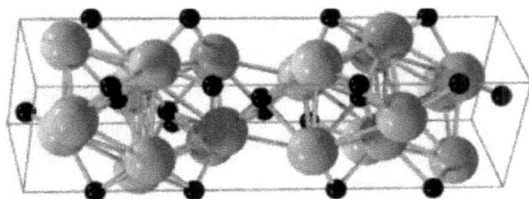

Fig. 1.2 Unit cell of α-aluminum oxide (corundum). Black spheres represent aluminum and gray spheres correspond to oxygen atoms. Reprinted from [Al-Abadleh and Grassian (2003)], Copyright 2003, with permission from Elsevier.

1.2.1.2 *Magnesium oxide*

Magnesium oxide or *magnesia* (MgO) has an energy gap of 7.8 eV and, similar to aluminum oxide, is widely used as a substrate for epitaxial growth of metal films and as a catalytic support. Magnesium oxide is a white solid mineral, which appears in nature in the form of *periclase*. The crystal structure of MgO is the same of NaCl. The MgO(001) surface is one of the most studied oxide surfaces due to its simple structure, stable stoichiometry, and easy preparation by cleavage. Furthermore, this surface is non-polar, which makes it a good model system for theoretical studies of insulating oxide surfaces.

1.2.1.3 *Silicon dioxide*

Silicon dioxide or *silica* (SiO_2) has a very large band gap of about 9.0 eV, which makes this material a superior electric insulator with high chemical stability. As well as being a major component in earth's crust, SiO_2 plays an important role in many technological applications, for example as a dielectric layer in microelectronics and as a catalyst support. One of the most common structures of silica is the *α-quartz*. This structure is formed by spirals of SiO_4 tetrahedra (Fig. 1.3) and is stable over a broad range of temperatures and pressures. In general, surfaces of α-quartz have very complicated structures, which makes their investigations on the nanoscale quite challenging.

1.2.2 *Mixed conducting oxide surfaces*

1.2.2.1 *Titanium dioxide*

Titanium dioxide or *titania* (TiO_2) has a band gap of 3.2 eV. Titanium dioxide has a wide range of applications. It is used in catalysis, in solar cells, as white pigment, as anti-corrosion coating, in ceramics, as well as in electric devices. Titanium dioxide usually crystallizes in the *rutile* structure, which consists of octahedra with a titanium atom in the center and oxygen atoms at the apices. *Anatase* and *brookite* are other known stable structures of titanium dioxide.

Bulk-truncated $TiO_2(110)$ surfaces (the most stable ones) reveal two kinds of termination: polar surfaces terminated with either Ti or O atoms, and a non-polar surface containing both undercoordinated Ti and O atoms. More precisely, on the non-polar surface coexist six-fold coordinated

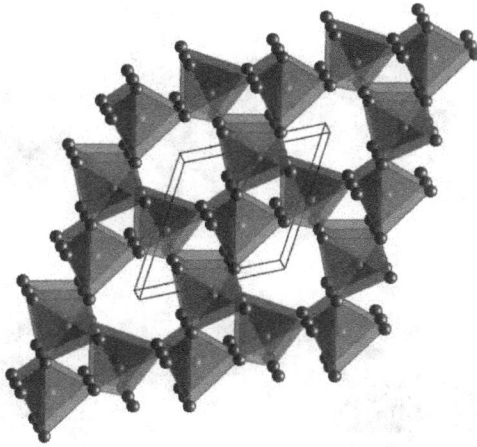

Fig. 1.3 Crystal structure of α-quartz. Dark spheres represent silicon and bright spheres correspond to oxygen atoms. Reproduced with permission from [de Graef and McHenry (2007)].

Ti atoms, five-fold coordinated Ti atoms, two-fold coordinated protruding O atoms (*bridging oxygens*), and fully three-fold coordinated O atoms (Fig. 1.4). Several experiments demonstrated that only the non-polar surface is stable. The (1×1) surface can undergo reconstructions upon annealing in UHV or under oxidizing conditions. The most commonly observed reconstruction is the (1×2) structure, which has been associated to wide rows added on every second Ti-exposed row of the (1×1) substrate [Takakusagi *et al.* (2003a)]. A comprehensive review of the surface properties of titanium dioxide has been given in [Diebold (2003)].

1.2.2.2 *Zinc oxide*

Zinc oxide (ZnO) has a band gap of 3.3 eV. Zinc oxide crystallizes in the hexagonal *wurtzite* structure, and occurs naturally as the mineral *zincite*. The wurtzite structure can be schematically described as a number of alternating planes formed by four-fold coordinated O^{2-} and Zn^{2+} ions and stacked along a common axis with alternating distances (Fig. 1.5). Cleaving the crystal perpendicular to the axis creates two polar surfaces (respectively Zn- and O-terminated) on the two sides of the cleavage, which have different physical and chemical properties. When doped, ZnO can be used to detect oxidizing and reducing gases. In such a context a crucial

(a)

(b)

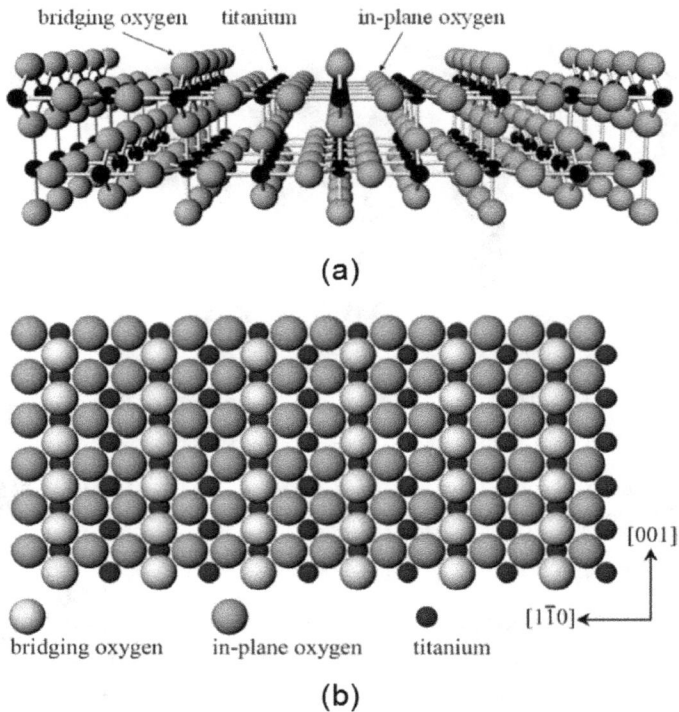

Fig. 1.4 (a) Ball-and-stick model of the rutile $TiO_2(110)$-(1×1) surface. (b) Two-dimensional model showing the stoichiometric surface structure. From [Bonnell and Garra (2008)]. Reproduced with permission from IOP Publishing.

role is played by the chemisorption of oxygen resulting in variations in the (low) conductivity of the oxide.

1.2.2.3 *Tin dioxide*

Tin dioxide (SnO_2) has a band gap of 3.6 eV and can be used as an oxidation catalyst or as a solid state gas sensing material. The mineral form of tin dioxide is called *cassiterite* and crystallizes in the rutile structure. The $SnO_2(110)$ surface can be easily reduced, resulting in a (1×2) surface reconstruction, which is not well understood. Oppositely to the (110) surface, $SnO_2(101)$ can form bulk truncations for both stoichiometric and reduced surfaces without undergoing complex reconstructions. These bulk truncations are shown in Fig. 1.6. A detailed discussion of tin dioxide surfaces can be found in [Batzill and Diebold (2005)].

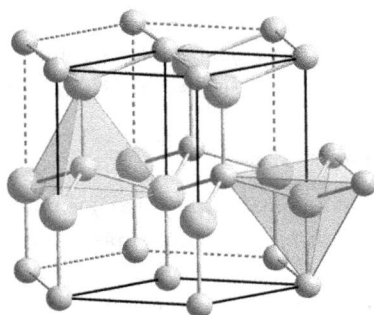

Fig. 1.5   Crystal structure of zinc oxide (wurtzite). Smaller spheres represent Zn^{2+} and larger spheres correspond to O^{2-} ions.

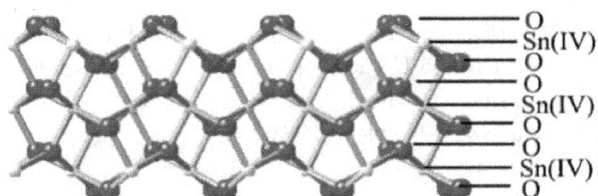

O
Sn(IV)
O
O
Sn(IV)
O
O
Sn(IV)
O

(a)

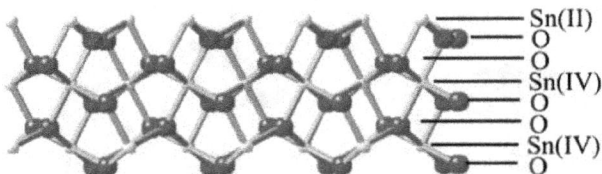

Sn(II)
O
O
Sn(IV)
O
O
Sn(IV)
O

(b)

Fig. 1.6   Ball-and-stick models of cross-sections through tin dioxide crystals with the (101) surface on top: (a) Stoichiometric bulk termination; (b) reduced surface with a Sn(II) surface layer. From [Batzill and Diebold (2007)]. Reproduced with permission of the PCCP Owner Societies.

1.2.2.4   *Cerium dioxide*

Cerium dioxide or *ceria* (CeO_2) has a relatively wide band gap of ca. 6 eV. Cerium dioxide crystallizes in the fluorite structure so that the thermodynamically most stable surface is $CeO_2(111)$. In the doped form CeO_2 is promising for solid oxide fuel cells due to its relatively high O^{2-}

ion conductivity in a wide range of temperatures. Cerium dioxide is also important in catalysis processes used to clean vehicle emissions since it can store and release oxygen in the exhaust stream of combustion engines.

1.2.2.5 *Strontium titanate*

Strontium titanate ($SrTiO_3$, band gap: 3.2 eV) is mainly used as a substrate for epitaxial growth of high-temperature superconductors and many oxide-based thin films. Strontium titanate crystallizes in the *perovskite* structure. At room temperature the cubic unit cell consists of a central Ti atom, which is octahedrally coordinated by six O atoms (Fig. 1.7). At the corners of the cube are located Sr atoms. This structure is essentially formed by a stack of alternating TiO_2 and SrO layers, so that a $SrTiO_3(001)$ surface may have either a TiO_2-termination or a SrO-termination. Many reconstructions of the $SrTiO_3(001)$ surface have been recognized, the interpretation of which is often unclear. Below 105 K the crystal structure of $SrTiO_3$ changes from cubic to tetragonal due the rotation of the oxygen octahedra around one of the main axes [Bruce and Cowley (1981)].

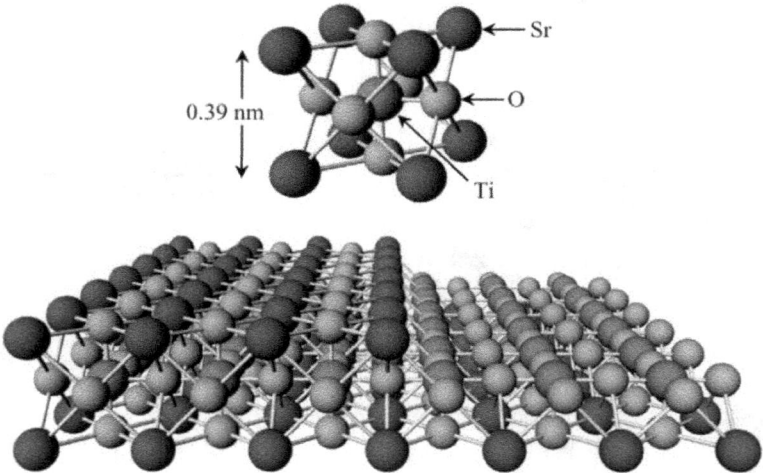

Fig. 1.7   Ball-and-stick model of the crystal structure of $SrTiO_3$ at room temperature showing the cubic unit cell and the (001) surface. This surface has two possible (1×1) bulk terminations, with SrO (left side) and TiO_2 (right side) stoichiometries. From [Bonnell and Garra (2008)]. Reproduced with permission from IOP Publishing.

Chapter 2

Preparation Techniques of Insulating Surfaces

In this chapter we introduce the techniques used to prepare model insulating crystal surfaces. Bulk-truncated surfaces can be obtained by cleavage in air or in ultra high vacuum, or by cutting and polishing in air, followed by cycles of sputtering and annealing in ultra high vacuum. Different procedures can be applied to create nanopatterns on these surfaces, such as evaporation spirals and nanofacets. Growing ultrathin films on a solid substrate is also a valuable method to create ordered structures of insulating materials. The film preparation may differ significantly for halide and oxide films, and the last ones may lose their insulating properties if their thickness is reduced to the nanometer range. We also mention how metals and large organic molecules on insulators can be deposited on insulators.

2.1 Ultra High Vacuum

In order to investigate clean structures, insulating surfaces have to be prepared and analyzed in *ultra high vacuum* (UHV) chambers, where the pressure is kept below 10^{-9} mbar. A typical setup, consisting of two chambers for preparation and analysis of the sample surfaces, is shown in Fig. 2.1. The vibrations of the instruments inside the analysis chamber can be minimized by springs and eddy-current damping stages. In such a way the motion of a probing tip close to the surface can be distinguished from the background noise and vertical resolution as low as 1 pm can be achieved by the methods of scanning probe microscopy described in Chapter 3.

To approach UHV conditions starting from atmospheric pressure a series of pumps has to be used, each of them operating in a specific pressure range. The final UHV stage can be achieved and maintained using turbomolecular, ion and titanium sublimation pumps. Pressures in this range are usually measured with ion gauges. More details can be found in surface science textbooks such as [Lüth (1996)] or [Oura *et al.* (2003)].

Fig. 2.1 A typical UHV setup to investigate insulating surfaces on the nanoscale: (1) Analysis chamber with STM/AFM; (2) preparation chamber; (3) valve separating the two chambers; (4) LEED/AES; (5) XPS; (6) sputter gun; (7) molecular evaporator with three Knudsen cell; (8) atom source. Courtesy of Prof. Ernst Meyer and Dr. Oliver Pfeiffer, University of Basel.

2.2 Preparation of Bulk Insulating Surfaces

2.2.1 *Halide surfaces*

Bulk alkali and alkaline earth halide surfaces can be easily cleaved by hitting a crystal against a rigid blade. Cleaving in air followed by *annealing* in UHV at 200-250 °C for a few hours usually results in clean atomically flat large terraces. However, some materials like CaF_2 are immediately contaminated after cleaving in air. As a result, while imaging these surfaces in UHV, granular structures may appear [Reichling *et al.* (1999)]. These structures are produced by chemical reactions between surface and air constituents, which irreversibly change the surface. Cleaving in UHV is the only way to avoid contamination, although it may cause another problem, i.e. the appearance of electric charges on and below the resulting surfaces. These charges create electrostatic forces on the probing tip, which can be much stronger than the forces responsible for topographic contrast in AFM imaging, i.e. the van der Waals forces (see Chapter 3). The strength of the charges induced by UHV cleavage depends on the materials, varies from cleavage to cleavage, and, as a matter of fact, it cannot be controlled. In order to reduce charging, the crystals have to be annealed after cleavage, usually at temperatures up to 150 °C. In such a case the sample has to be mechanically fixed in a sample holder, which does not outgas at high temperatures.

2.2.2 *Oxide surfaces*

Cleavage in UHV is the simplest way to prepare clean and well-ordered single crystal oxide surfaces. The resulting surfaces are generally stoichiometric and exhibit a low defect density. However, this method has two severe limitations. First, cleavage can be only performed on brittle materials, like MgO or ZnO crystals. Second, cleavage is only feasible along well-defined crystallographic directions corresponding to the formation of low surface-energy surfaces. Crystals with the rock salt structure such as MgO can be cleaved along the (001) surfaces, while crystals with the wurtzite structure such as ZnO can be cleaved along the non-polar faces (1010), although successful cleavages of some polar surfaces, such as MgO(111), ZnO(0001), and ZnO(000$\bar{1}$), have been also reported [Henry (1998); Freund *et al.* (1996)].

Oxide surfaces can be also prepared by mechanical *cutting* and *polishing* in air. This method is applied to the most common oxide crystals, such as SiO_2, TiO_2, Al_2O_3, and $SrTiO_3$, as well as to the otherwise cleavable MgO and ZnO crystals. In such a way different orientation surfaces can be obtained. As a next step, *ion sputtering* is required to remove surface impurities due to the polishing process, air contamination or segregation of bulk impurities to the surface. Finally, annealing in oxygen or vacuum should be used to get clean and well-ordered surfaces. Quite often several cycles of sputtering and annealing are necessary to achieve adequate surface ordering.

2.2.3   *Nanostructuring of insulating surfaces*

Insulating crystal surfaces can be patterned in several ways to produce templates where adsorbates like metal clusters or organic molecules can accommodate and form ordered structures. Here we discuss patterns created by heating at high temperatures or by slightly miscutting insulating surfaces. Other structures produced by photon or electron irradiation and by ion bombardment are introduced in Chapters 7 and 8, while patterns induced by chemical impurities or caused by nanoindentation and abrasion processes are considered in Chapters 4 and 12.

2.2.3.1   *Evaporation spirals on alkali halides*

Heating of alkali halide crystals in UHV results in molecular evaporation from corner sites of cleavage steps. This process smooths the steps and results in the formation of *evaporation spirals* around the intersection of dislocations with the surface [Bethge (1964); Yamamoto *et al.* (1989); Nony *et al.* (2004a)]. The evaporation spirals form regular arrays of monatomic or diatomic steps, which separate atomically flat terraces. Typical terrace widths are between 50 and 150 nm [Fig. 2.2(a)]. The enhanced interaction at the steps of the spirals makes them good candidates for guided growth of adsorbates, as demonstrated in Chapters 9 and 10, and is also responsible for increased friction at such locations (Chapter 12).

2.2.3.2   *Faceting of halide and oxide surfaces*

As repeatedly pointed out, certain surfaces of crystalline solids are more stable than others. Heating less stable surfaces at high temperatures enhances the mobility or desorption of atoms or molecules and, as a consequence, *faceting* may occur, resulting in the creation of more stable surfaces. Faceted surfaces can exhibit very regular structures on a long range. An example is given in Fig. 2.2(b), where a ridge-valley pattern created on a NaCl(110) surface is shown [Sugawara *et al.* (1997a)]. The two slopes correspond to (100) and (010) facets, as revealed by iron clusters deposited on the pattern. The (110) surface was obtained by short etching in water before transferring into the vacuum chamber and annealing. Arrays of pyramidal structures were also reported after annealing a NaCl(111) surface [Sugawara *et al.* (1997b)].

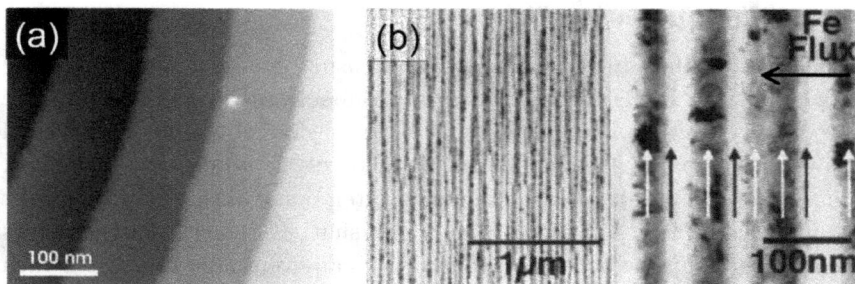

Fig. 2.2 (a) AFM image of a KBr(001) surface showing terraces separated by regularly spaced monatomic steps, which are part of an evaporation spiral. The surface has been previously heated at 380 °C for 20 min. From [Nony *et al.* (2004a)]. Reproduced with permission from IOP Publishing. (b) Left: TEM image of an array of Fe nanowires grown on a silicon layer deposited on a stepped NaCl(110) substrate. The periodicity of the pattern is about 90 nm. Right: The white arrows indicate thick lines due to shadow deposition. The black arrows show lines of ultrafine particles that nucleated at the bottom of the grooves and remained there after formation of the wires. Reused with permission from [Sugawara *et al.* (1997a)]. Copyright 1997, American Institute of Physics.

Faceting of oxide surfaces has been first demonstrated on the (110) and (111) surfaces of MgO, which were found to form (100) facets upon annealing [Henrich (1976)]. Annealing a ($10\bar{1}0$) surface of Al_2O_3 at 1400 °C resulted in the formation of parallel ridges and valleys [Heffelfinger and Carter (1997)]. Here one slope corresponded to a ($1\bar{1}02$) facet, and the other slope was a complex surface not resembling any simple crystallographic plane. After several hours of annealing the distance between adjacent ridges could be stabilized at approximately 300 nm.

2.3 Deposition of Insulating Films, Metals and Organic Molecules

Growing thin films of an insulating material on a solid substrate is another important method to prepare ordered surfaces. Depending on the film thickness, the film surface can be strongly influenced by the underlying substrate or reveal the same properties of bulk-truncated surfaces. Metal adsorbates and large organic molecules can also be deposited on insulating surfaces, which usually results in the formation of metal clusters and ordered molecular islands.

2.3.1 *Thin insulating films*

Alkali halide films are usually prepared using *molecular beam epitaxy* (MBE). In the MBE technique, pure precursors are heated in separate evaporators (*Knudsen cells*) until they begin to slowly sublimate. A typical Knudsen cell contains a crucible (made of pyrolytic boron nitride, quartz, tungsten or graphite), heating filaments (often made of tantalum), a water cooling system, heat shields and an orifice shutter. The gaseous elements then condense on the substrate, where they can eventually react with each other. The term 'beam' means that the evaporated molecules do not interact with each other or with gases in the vacuum chamber until they reach the substrate, due to their long mean free paths. The evaporation rate can be calibrated using a *quartz crystal microbalance* (QCM) with an accuracy of better than 10%.

Alkali halide molecules evaporate in significant rates from edge and corner sites at temperatures as low as two thirds of the melting temperature of the crystal. As a consequence, heating a powder of an alkali halide in a Knudsen cell allows growing stoichiometric films in a relatively simple way. The strong ionic cohesion within alkali halide films results in the formation of large islands (often overgrowing substrate steps) with exceptional atomic order. The structures of these islands are examined in Chapter 5.

Oxide films can be prepared by direct vapor evaporation of bulk oxides using a high energy laser or thermal heating. Alternatively, metals can be evaporated onto a clean surface in an oxygen atmosphere, which leads to the formation of an oxide layer. This procedure is used, for instance, in the case of MgO. Alternatively, single crystal surfaces of metals and alloys can be oxidized. This technique has been successfully adopted for preparing thin films of SiO_2 and Al_2O_3. Ultrathin oxide films can reveal significant conductivity when grown on a conducting surface. In such a case, the film structure can be investigated by scanning tunneling microscopy and other surface science techniques, even when this is not the case for the corresponding bulk-truncated surfaces. The structures of thin oxide films, as resolved by SPM, are discussed in Chapter 5 as well.

2.3.2 *Metal adsorbates on insulators*

The most common way to deposit metal overlayers onto a solid surface is *physical vapor deposition* (PVD) under UHV conditions. When a metal vapor is deposited onto a substrate, various atomic processes take place. These processes are schematically shown in Fig. 2.3, where the corresponding characteristic energies per adatom are also indicated. The interface formation is basically guided by three atomic processes: in-plane surface atom diffusion, out-of-plane surface atom diffusion (including down-step diffusion and up-step diffusion) and atom interdiffusion. The energies shown in Fig. 2.3 are strongly dependent on the metalsubstrate interaction. If the interaction between substrate and metal atoms is stronger than between neighboring metals atoms the growth occurs in a *layer-by-layer* mode. In the opposite case, three-dimensional clusters or islands are formed (*Volmer-Weber* mode). The latter scenario is the most common on insulating substrates, as shown in Chapter 9.

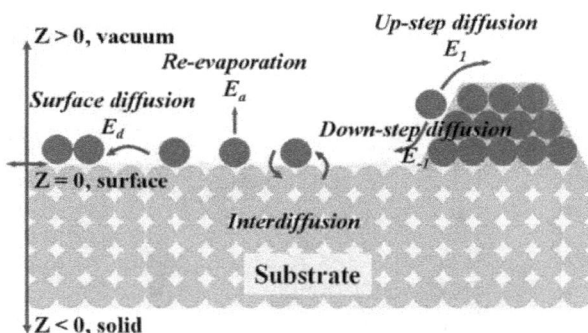

Fig. 2.3 Elementary processes taking place when metal atoms are deposited onto a solid surface. E_a is the adsorption energy of an atom, E_d the surface diffusion energy, E_1 the up-step diffusion energy, and E_{-1} the down-step diffusion energy. Reprinted from [Fu and Wagner (2007)]. Copyright 2007, with permission from Elsevier.

Metal overlayers can be also produced by *chemical vapor deposition* (CVD). In such a case organometallic precursors are exposed to the substrate in vacuum, and ligands are subsequently removed by heating.

2.3.3 *Organic molecules on insulators*

Organized molecular arrays are usually grown on insulating substrates by two different procedures. Monolayers of amphiphilic molecules with a hydrophilic head and a hydrophobic tail, such as fatty acids, can be deposited from the surface of a liquid onto a solid by immersing and emersing the solid substrate into or from the liquid. A molecular layer is added at each step, resulting in the formation of a *Langmuir-Blodgett film* with well defined thickness. Alternatively, certain molecules can be designed to adsorb spontaneously on the surfaces of compatible solids from vapor phases. As a result ordered organic layers *self-assemble* on the support. This technique is usually preferred in UHV studies.

In organic-inorganic heteroepitaxy different types of growth have been classified [Hooks *et al.* (2001)]. An important result is that, for a large number of systems, the molecular layers do not grow in a commensurate fashion. Instead, the overlayers adopt so-called *point-on-line coincident* structures in which all molecules are placed on primitive lattice lines of the substrate. The sketch in Fig. 2.4 gives an example for a point-on-line coincident unit cell on a conductive graphite surface. Understanding whether similar pictures apply also to insulating substrates is being currently investigated in challenging SPM studies (see Chapter 10).

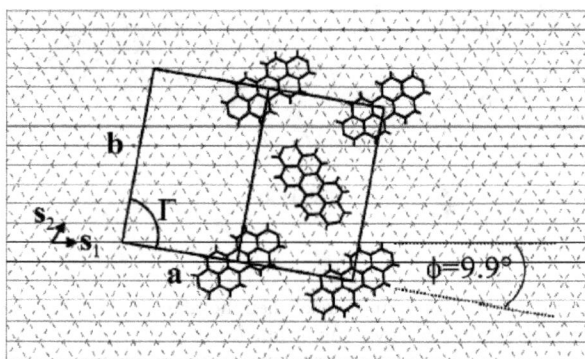

Fig. 2.4 Schematic arrangement of a unit cell of PTCDA molecules (Sec. 10.1) on graphite. Note that the lattice vectors of the adsorbate (**a** and **b**) do not coincide with lattice points of the graphite substrate, but the unit cell is oriented such that the lattice vectors end on lattice lines (parallel to s_1). Reused with permission from [Hoshino *et al.* (1994)]. Copyright 1994, American Institute of Physics.

Chapter 3

Scanning Probe Microscopy in Ultra High Vacuum

The establishment of *scanning probe microscopy* (SPM) opened the possibility to study the structure of surfaces with high resolution in real space. Atomic force microscopy and scanning tunneling microscopy rely on the fact that forces and, respectively, charge flow between atoms are very sensitive to the physical environment. Here we mainly focus on AFM, which is the most important technique for imaging insulating surfaces on the nanoscale. Scanning tunneling microscopy can be applied to mixed conducting oxide surfaces and thin insulating films, and is also briefly discussed. Atomic force microscopy is based on a delicate balance of chemical, van der Waals and electrostatic forces acting between probe and sample surface. Thus, understanding and reproducing the imaging process requires a considerable theoretical effort, as outlined in the last part of the chapter. Other methods of surface characterization applicable to insulating surfaces such as low-energy electron diffraction, electron spectroscopies or helium scattering are described in standard textbooks on surface science and will not be considered here.

3.1 Atomic Force Microscopy

In a typical *atomic force microscope* (AFM) [Binnig *et al.* (1986)] a sharp micro-fabricated tip is scanned over a surface. The tip is attached to a cantilever force sensor, the sensitivity of which can be well below 1 nN. Images of the surface topography are recorded by controlling the tip-sample distance in order to maintain a constant (*normal*) force. This is made possible by using piezoresistive cantilevers, or, most commonly, a light beam reflected from the back side of the cantilever into a photodetector, which allows to monitor the cantilever bending (Fig. 3.1). The (*lateral*) friction force between tip and surface is responsible for the cantilever torsion and can be monitored if the photodetector is equipped with four quadrants. If this is the case the AFM can be used as a *friction force microscope* (FFM). A home-built AFM, optimized for friction measurements in UHV, is represented in Fig. 3.2.

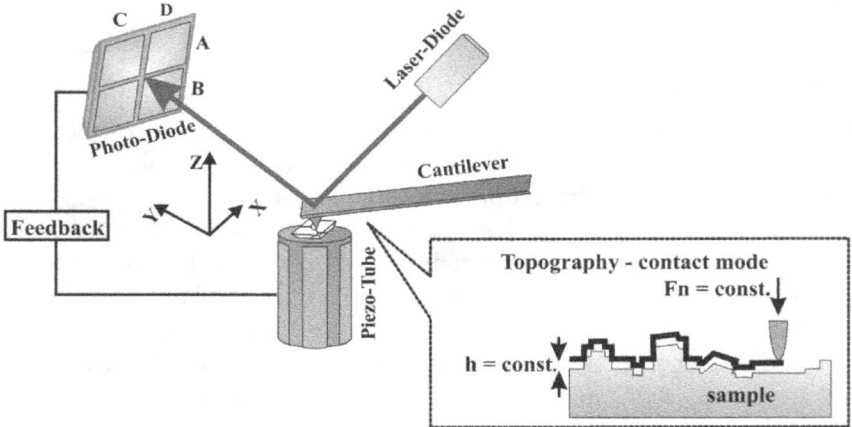

Fig. 3.1 Schematic view of an atomic force microscope (AFM). The inset shows the principle of operation in contact mode.

The tip-sample force can be related not only to the static bending or torsion of the cantilever. In *dynamic AFM* techniques [Meyer *et al.* (2004)], the cantilever is excited in the vicinity of one of its resonance frequencies, and the tip-sample interaction is estimated from the oscillation amplitude or the shift of the resonance frequency. In such a way sliding contact between tip and sample is avoided, and damage to tip and surface is considerably reduced.

The imaging process in AFM takes place continuously above the surface. The tip is usually scanned at a constant velocity forwards and backwards in the so-called 'fast' scan direction, then the motion is stopped, the tip is shifted by a short distance in the 'slow' scan direction, which is perpendicular to the fast scan direction, and the process is repeated several times to produce a two-dimensional map.

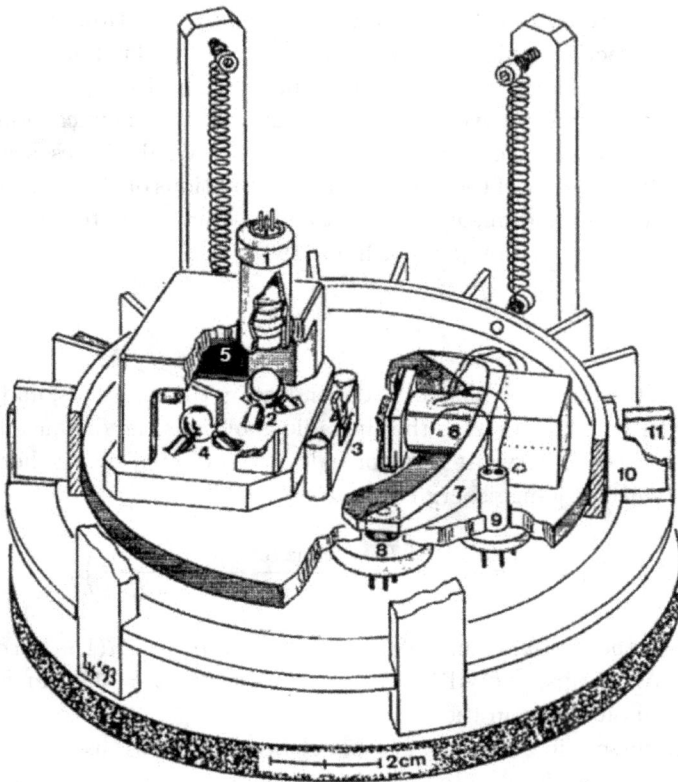

Fig. 3.2 Sketch of the home-built UHV-AFM in the analysis chamber in Fig. 2.1: (1) light source with optics; (2), (4) plane mirrors on spherical stepping motors; (3) cantilever holder with probing tip; (5) quadrant photodiode; (6) tube scanner with sample; (7) slider of two-dimensional stepping motor; (8) driving piezo; (9) fixed post; (10), (11) eddy current damping for spring suspension. Courtesy of Prof. Ernst Meyer, University of Basel.

3.1.1 *Relevant forces in AFM*

The most important interactions in atomic force microscopy are due to short-range *chemical forces*. Chemical forces are sensitive to single atoms and are responsible for atomic resolution. They also define the atomic structure of the tip and surface, and cause atomic displacements when the tip is brought in close proximity to the surface.

The *van der Waals* (vdW) *forces* are due to the electromagnetic interaction of fluctuating dipoles in the atoms forming tip and surface.

These forces are extremely weak on the atomic level. However, they are generally attractive, which can result in forces of several nanonewtons when small interactions between individual atoms of tip and sample are summed up. In such a way the vdW forces can exceed the chemical forces and dominate the tip-surface interaction. Van der Waals forces are always present independently of the tip and surface conditions or the environmental conditions of the experiment. In the case of a sphere close to a flat surface, the vdW force is given by [Israelachvili (1991)]

$$F_{vdW}(z) = \frac{HR}{6z^2},$$

where H is the *Hamaker constant* (dependent on the materials, and usually in the order of 10^{-19} J), R is the tip radius, and z is the distance between tip and surface. In the case of a conical tip terminated by a spherical cap and a flat surface [Guggisberg *et al.* (2000)]:

$$F_{VdW}(z) = -\frac{H}{6} \left(\frac{R}{z^2} + \frac{\tan^2 \alpha}{z + R_\alpha} - \frac{R_\alpha}{z(z + R_\alpha)} \right),$$

where α is the half-aperture of the cone, and $R_\alpha = R(1 - \sin \alpha)$. For standard AFM tips, the vdW forces at a distance $z = 0.5$ nm from the surface turn out to be in the nanonewton range.

As discussed in Chapter 2, the cleavage process used to prepare atomically flat surfaces often results in charges trapped at the sample surface. Other surface preparation techniques such as ion sputtering in UHV may also lead to charging effects. If localized charges are also present at the tip apex, *electrostatic forces* are generated, the strength and distance dependence of which is given by the Coulomb's law. Electrostatic forces also act between charged surfaces and conductive tips. Considering the tip-surface system as a capacitor with a distance-dependent capacitance C, the electrostatic forces are given by

$$F_{el} = \frac{\partial C}{\partial z} (V_{bias} - V_{cpd})^2,$$

where V_{bias} is the (*bias*) voltage applied between tip and surface and V_{cpd} is the contact potential difference produced by the different work functions of tip and surface.

Other contributions to the tip-surface interaction such as magnetic forces and capillary forces are not considered in this book.

3.1.2 Contact AFM

In *contact AFM*, the static deflection of the cantilever is used as a feedback signal (Fig. 3.1). When the probing tip is scanned in contact with the sample surface, both the cantilever and the contact region are elastically deformed. Rectangular cantilevers undergo both normal bending and torsion. The bending spring constant is given by $c_B = Ewt^3/4l^3$, where E is the Young modulus, and l, w and t are respectively the length, width and thickness of the cantilever. The torsional spring constant is $c_T = Gwt^3/3h^2l$, where G is the shear modulus and h is the tip height. Typical values for commercial silicon cantilevers are $c_B \sim 0.1$ N/m and $c_T \sim 100$ N/m. These quantities combine with the normal and lateral stiffness of the contact region, respectively $k_{N,\text{con}}$ and $k_{L,\text{con}}$, to give the effective spring constants of the sliding system, $k_{N,\text{eff}}$ and $k_{L,\text{eff}}$, which are required to calibrate the normal and lateral forces acting on the AFM tip. Approximate relations can be obtained by considering the cantilever and the contact region as a series of springs, as in Fig. 3.3:

$$\frac{1}{k_{(N,L),\text{eff}}} = \frac{1}{k_{(N,L),\text{lev}}} + \frac{1}{k_{(N,L),\text{con}}}.$$

In the previous formulas, $k_{N,\text{lev}} = c_B$ and $k_{L,\text{lev}} = (3/2)c_T(h/l)$ [Lüthi *et al.* (1995)], and the tip is assumed infinitely stiff. When contact AFM is performed on insulating crystal surfaces, the inequality $k_{N,\text{lev}} \ll k_{N,\text{con}}$ usually holds, and any normal force applied on the tip mainly results in cantilever bending [Carpick *et al.* (1997)]. On alkali halide surfaces it has been recognized that $k_{L,\text{con}} \ll k_{L,\text{lev}}$ (Sec. 12.2) so that the lateral force is dominated by the frictional forces arising in the contact region. However, this conclusion seems to be rather due to strong deformations in the tip apex than in the ionic crystal [Krylov *et al.* (2006)].

3.1.3 Non-contact AFM

In *non-contact AFM* (NC-AFM) the probing tip oscillates with an amplitude of a few nanometers at the bending resonance frequency of the cantilever. The oscillation is usually applied by a piezoactuator mounted at the base of the cantilever itself. Silicon cantilevers with normal spring constants k_N of few tens of N/m and resonance frequencies f_0 in the order of 10^5 Hz are typically used, with corresponding Q factors $\sim 10^4$ (in UHV).

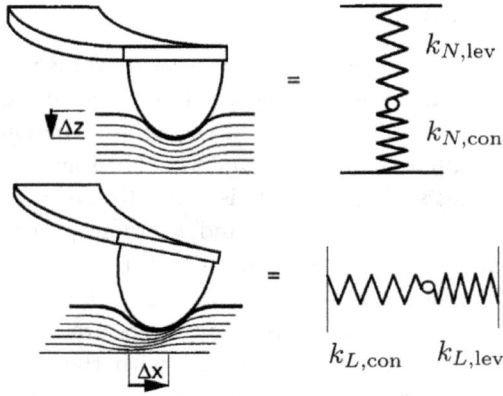

Fig. 3.3 Schematic picture of the contact formed by an AFM tip sliding on an elastic surface. Reused with permission from [Carpick *et al.* (1997)]. Copyright 1997, American Institute of Physics.

In NC-AFM the system formed by cantilever and tip can be represented as a damped harmonic oscillator [Meyer *et al.* (2004)]:

$$m\ddot{z} = -k_N \left[z - A_{\text{exc}} \cos(\omega t + \varphi) \right] - \gamma \dot{z} + F(z), \qquad (3.1)$$

where m is the effective mass of the cantilever, z is the vertical position of the tip, γ is a damping coefficient (related to the internal friction of the material), and $F(z)$ is the force between tip and sample surface. For rectangular cantilevers m is approximately one fourth of the cantilever mass [Gürgöze (2005)]. The excitation has an amplitude A_{exc} and the response of the cantilever occurs with a phase shift φ with respect to the excitation signal. Since the (normal) friction force $-\gamma \dot{z}$ is compensated by the driving force $F_{\text{exc}} = k A_{\text{exc}} \cos(\omega t + \varphi)$, equation (3.1) simply becomes

$$m\ddot{z} = -kz + F(z). \qquad (3.2)$$

The excitation needed to maintain the oscillation amplitude of the cantilever results in the so-called *damping signal* of the NC-AFM.

If the tip oscillations were small compared to the characteristic decay length of the force $F(z)$, a linear expansion of $F(z)$ would be possible, leading to a proportionality relation between the *shift* Δf of the resonance frequency and the force gradient in the z direction. However, this is usually not the case in the NC-AFM experiments performed in UHV. Here the interaction between tip and sample modifies the harmonic motion of the cantilever only close to the lower turning point of the tip. Assuming that the tip undergoes harmonic oscillations of amplitude A, and that the frequency shift Δf is small, the following relation between measurable parameters and the force F averaged over the oscillation cycle has been derived [Giessibl (1997)]:

$$\Delta f = \frac{f_0}{\pi k_N A} \int_0^{2\pi/\omega} \sin \omega t F(z_0 + A \sin \omega t)\, dt. \qquad (3.3)$$

The integral on the right-hand-side of (3.3) can be evaluated for large oscillation amplitudes, assuming different force-distance relations. If the distance between tip and surface at closest approach is smaller than the tip radius, the long-range interactions are dominated by the spherical cap of the tip. In this case simple expressions for the frequency-shift Δf_{el} due to the electrostatic interaction and the frequency shift Δf_{vdW} due to the vdW interaction have been obtained [Guggisberg *et al.* (2000)]:

$$\Delta f_{el} = -\frac{f_0}{k_N A^{3/2}} \cdot \frac{\pi \varepsilon_0 R (V_{bias} - V_{cpd})^2}{\sqrt{2 z_{min}}}, \qquad (3.4)$$

$$\Delta f_{vdW} = -\frac{f_0}{k_N A^{3/2}} \cdot \frac{HR}{6 s \sqrt{2 z_{min}}}.$$

In the formulas (3.4), z_{min} is the closest distance between the surface and the mesoscopic part of the tip, and the other parameters are defined as in Sec. 3.1.1.

It is worth to note that the force vs. distance curves $F(z)$ can be reconstructed from frequency shift vs. distance curves $\Delta f(z)$ without any assumptions for the force law. For such purpose an iterative method has been developed [Dürig (2000)]. Various examples of *force spectroscopy* on insulators are given in Chapter 11.

3.1.3.1 *Tuning fork sensors*

Tuning fork sensors have been introduced to improve stability and sensitivity of NC-AFM [Giessibl *et al.* (2004)]. Fig. 3.4 shows a tuning fork sensor formed by a tip attached to the end of a large quartz cantilever. This system has a very large spring constant (in the order of 1 kN/m), oscillates with a very small amplitude in the subnanometer range, and has a relatively small resonance frequency of a few tens of kHz. Tuning fork sensors are promising in nanomanipulation experiments on insulating surfaces, as discussed in Chapter 13.

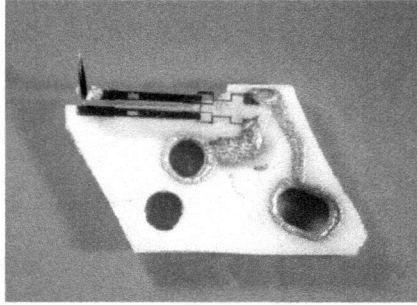

Fig. 3.4 Micrograph of a tuning fork sensor. The tuning fork is glued onto an alumina substrate. A probing tungsten tip is mounted at one 'leg' of the fork. Reprinted with permission from [Giessibl (2003)]. Copyright (2003) by the American Physical Society.

3.1.4 *Kelvin probe force microscopy*

According to (3.4) the frequency shift in NC-AFM depends quadratically on the difference between the bias voltage V_{bias} and the contact potential difference V_{cpd} between tip and surface, so that V_{cpd} can be determined by recording Δf as a function of the bias V_{bias}. Alternatively, the so-called *Kelvin probe force microscopy* (KPFM) technique can be used for a continuous mapping of V_{cpd}. In KPFM, the bias voltage is modulated by a small AC voltage: $V_{bias} = V_{DC} + V_{AC} \sin \omega t$. If the modulation frequency ω is low enough, the frequency shift Δf has two harmonic components:

$$\Delta f_\omega \propto (V_{DC} - V_{cpd}) V_{AC} \sin \omega t,$$

and

$$\Delta f_{2\omega} \propto V_{AC}^2 \cos 2\omega t.$$

In a feedback circuit, V_{DC} can be adjusted to maintain $\Delta f_\omega = 0$. The corresponding maps of V_{DC} (*potential images*) reproduce the distribution of the contact potential difference V_{cpd} between tip and surface. Examples of potential images on insulating surfaces are given in Chapters 4 and 9.

3.2 Scanning Tunneling Microscopy

3.2.1 *Scanning tunneling microscopy*

In a *scanning tunneling microscope* (STM) [Binnig *et al.* (1982, 1983)] piezoelectric transducers bring a sharp metallic tip down to a distance of a few Å from a conducting surface (Fig. 3.5), where the wave functions of the electrons of tip and surface overlap. A bias voltage, V_{bias}, applied between tip and sample causes electrons to tunnel from the tip to the surface or vice versa, depending on the sign. The resulting tunneling current, I_t, can range from a fraction of pA to a few nA depending on the materials, distance and bias voltage.

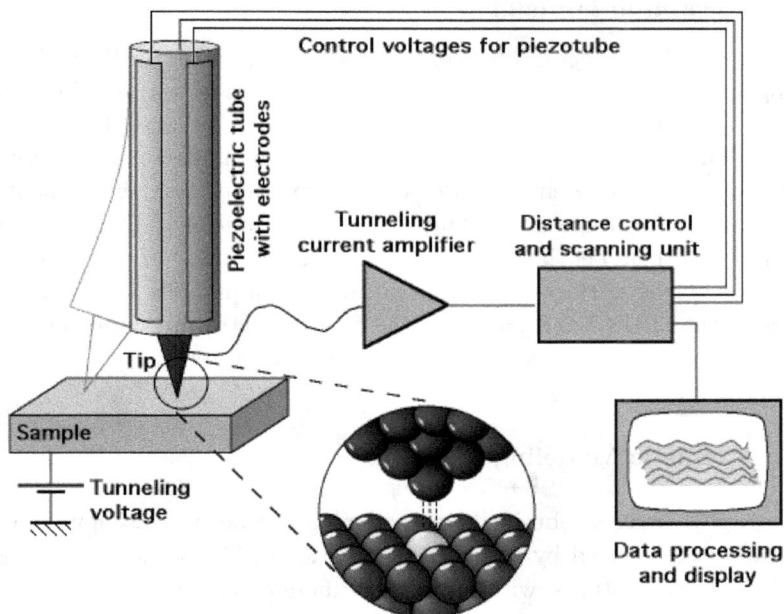

Fig. 3.5   Schematic view of a scanning tunneling microscope (STM).

As a first approximation, the tunneling current decays exponentially with the tip-sample distance z:

$$I_t(z) \propto V_{\text{bias}} \rho_S(E_F) e^{-2\kappa z}. \qquad (3.5)$$

In this formula $\rho_S(E_F)$ is the *local density of states* (LDOS) at the Fermi level and $\kappa = \sqrt{2m\Phi/\hbar^2}$, where Φ is the height of the tunneling barrier. If the current I_t is kept constant by a feedback loop while scanning, a constant charge density surface can be mapped.

At first sight, STM imaging seems to be not feasible on insulating surfaces. However, in the case of mixed conducting oxides, the conductivity of the sample can be increased by annealing at very high temperatures. Furthermore, thin insulating films on a conductive substrate can be imaged by STM if their thickness does not exceed a few monolayers, so that tip and substrate are electrically coupled. Successful applications of STM imaging to these systems are presented in Chapters 4 and 5 respectively.

3.2.2   *Scanning tunneling spectroscopy*

In *scanning tunneling spectroscopy* (STS), the current feedback loop is disengaged and the probing tip is positioned at a given location above the surface. The bias voltage V_{bias} is then swept between two extreme values. The *I-V curve* so produced can be differentiated numerically, or, alternatively, the differential can be directly extracted from experiments in which the bias voltage is oscillated with low frequency. The differential dI/dV curve can be then compared with simulations of the electronic structure of the surface. On thin insulating films STS can reveal the development of the band gap with the film thickness, and also probe the electronic structure of adsorbates, as shown in Chapter 11.

3.3   Atomistic Modeling of SPM

Topography images obtained by scanning probe microscopy can be accurately reproduced by computer simulations. The modeling process is divided in several stages, with important differences between NC-AFM and STM (contact mode AFM is discussed separately in Chapter 12). Assuming that the surface structure is known, a realistic tip model has first to be established, based on experimental or theoretical results. For AFM, the macroscopic shape of the tip is essential, since it determines the long-range interactions with the sample surface. This is not the case for STM, where imaging mainly results from the microscopic structure of the tip and the electronic states, and from the atom configuration at the tip apex. As a next step, the interactions between tip and surface are modeled, and the cantilever oscillations or the current resulting from these interactions are calculated. Finally, a theoretical image is built up and compared with experimental results.

On the micrometer scale, AFM tips have usually a pyramidal shape. However, no direct methods are available for imaging the tip apex (the so-called *nanotip*). Tips are usually oxidized due to exposure to the atmosphere and, although the oxide layer can be removed by ion sputtering, they can still be contaminated by the residual water always present in a UHV chamber. Furthermore, tip crashes may occur while imaging, resulting in material transfer onto the nanotip. Crashes are due to enhanced tip-surface interaction and debris on the surface, or they can also be intentionally provoked to improve the resolution. For these reasons only idealized nanotips can be used or, alternatively, a series of tips has to be modeled and compared to experimental results in order to recognize the features which better reproduce the data.

The tip-surface interaction can be divided into the chemical forces and the macroscopic forces acting between tip and surface. The last ones always include the vdW and electrostatic interactions, discussed in Sec. 3.1.1. Most simulations have suggested that only the short-range chemical forces are responsible for atomic resolution [Hofer *et al.* (2003)], whereas the macroscopic forces can be considered as an attractive force acting in the background. This background is important to reproduce frequency shifts in NC-AFM, but it does not depend on the atomic configuration at the tip apex and does not result in atomic displacements. Thus, the macroscopic forces can be neglected in simulations of STM experiments with high resolution. In AFM modeling the chemical and macroscopic forces can be calculated separately and combined to give the total force between tip and surface.

Figure 3.6 shows a MgO nanotip and a LiF surface used as model system for NC-AFM simulations [Livshits *et al.* (1999)]. The nanotip and the upper layers of the surface are divided into two regions. The ions in the first region are allowed to relax, while the ions in the second region are kept fixed in order to reproduce the potential of the bulk lattice and of the remaining tip ions. Provided that the unit cell of the surface is large enough, periodic boundary conditions can be imposed to simulate an infinite surface. The total force acting on the tip can be calculated at different tip-surface distances. A map of the microscopic tip-surface interaction can be generated by repeating the calculation over many surface locations. Combined with the macroscopic force, the total force can be thus estimated as a function of the relative position of tip and surface.

Maps of tunneling currents in STM can be easily simulated by shifting the probing tip laterally and vertically across the surface. Since the tunneling conditions define a constant current contour, no parameters are required to fit the experimental data. In the case of NC-AFM, once the three-dimensional force field above the surface is known, the trajectory of cantilever and tip can be determined using integration algorithms with time steps much smaller than the oscillation period of the cantilever. The values of the resonance frequency f_0, oscillation amplitude A, spring constant k_N, and frequency shift Δf can be estimated from the experimental data, whereas other parameters, such as the tip radius, are based on realistic assumptions. If the time step of the simulations is much larger than the time scale of the atomic relaxation processes (in the order of $10^{-12} - 10^{-13}$ s) the force field can be instantaneously changed whenever a structural change on the atomic scale occurs. In such a way, not only imaging but also manipulation processes can be reproduced, as shown in Chapter 13.

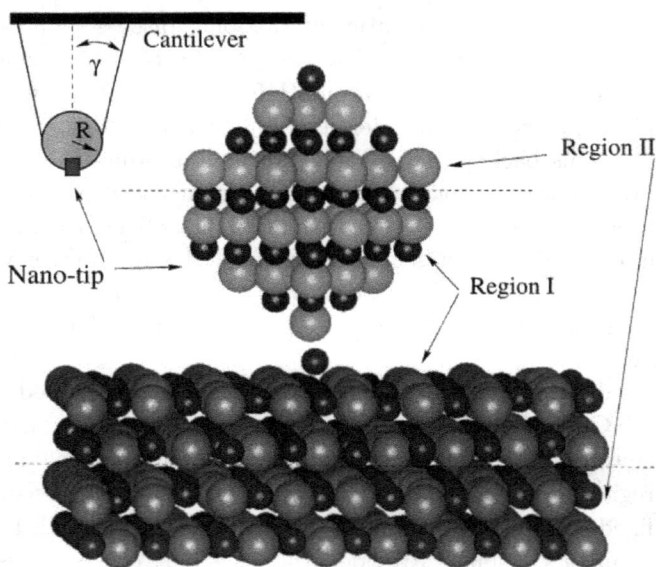

Fig. 3.6 Schematic view of an AFM tip over an alkali halide surface. A possible atomic configuration of the MgO nanotip is also shown. Reprinted with permission from [Livshits *et al.* (1999)]. Copyright (1999) by the American Physical Society.

Chapter 4

Scanning Probe Microscopy on Bulk Insulating Surfaces

In this chapter we discuss how atomic force microscopy has been applied to image bulk-truncated halide and oxide surfaces, prepared by the methods described in Chapter 2. After the achievement of atomic resolution on halide surfaces, several groups have succeeded in recognizing different kinds of nanostructures, ranging from single point defects to more 'exotic' Suzuki phases. Atomic scale features on oxide surfaces are more difficult to resolve, and only few successful investigations have been reported based on AFM. Simulations of NC-AFM images on insulating surfaces obtained by the method introduced in Sec. 3.3 are also presented. The excellent agreement with the experiments shows that the interactions between probing tips and insulating surfaces are well understood, which is promising for further investigations on other systems.

4.1 Halide Surfaces

4.1.1 *Alkali halide surfaces*

The first atomically resolved images of bulk-truncated alkali halide surfaces were acquired on (001) surfaces of NaF, RbBr, LiF, KI and NaCl using NC-AFM at room temperature (Fig. 4.1) [Bammerlin *et al.* (1998)]. Atomic resolution was also reported on KBr(001) at very low temperature (7 K) [Hoffmann *et al.* (2002)]. In all cases, the lattice period in the AFM images corresponded to the bulk lattice of equally charged ions, indicating that only ions of one kind are resolved as protrusions by NC-AFM. An interesting investigation focused on a *mixed crystal* containing 60% KCl and 40% KBr [Bennewitz *et al.* (2002)]. Since the components of these solid solutions are usually randomly mixed on the atomic scale, the crystal was imaged without further annealing after UHV cleavage to avoid any possible reorganization of the surface. Atomically resolved topographies showed strongly varying depressions, which were associated to the Cl^- and Br^- ions (Fig. 4.2). Since depressions correspond to sites of reduced attractive force or repulsive electrostatic potential, the authors concluded that the tip apex was carrying a negative charge. A further series of measurements revealed atomic contrast also in the damping signal, which appeared enhanced above certain protrusions [Bennewitz *et al.* (2004)]. Comparing these protrusions with the relative density of the anions, the enhanced damping could be attributed to Br^- sites.

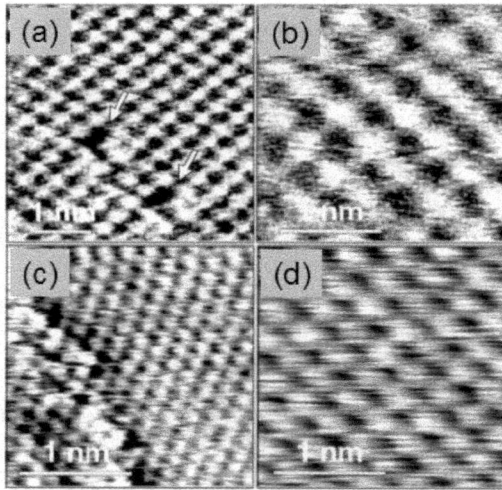

Fig. 4.1 NC-AFM images of (a) NaCl(001), (b) NaF(001), (c) LiF(001) and (d) RbBr(001) surfaces. Marked spots in (a) represent atomic scale defects, while the defects in (c) presumably correspond to Mg clusters. Reprinted from [Bammerlin *et al.* (1998)]. With kind permission from Springer Science+Business Media.

The (001) surfaces of NaCl and KCl were also studied by KPFM [Barth and Henry (2006c, 2007)]. Here, bright round patches were found in correspondence of corner and kink sites (Fig. 4.3). From the sign of the voltage V_{cpd} in the potential images, the authors concluded that the kinks were negatively charged. The reason for that was attributed to the so-called *surface double layer* mechanism. [Frenkel (1946)]. At room temperature divalent impurities such as Ca^{2+}, which are commonly found in alkali halide crystals, require the creation of cation vacancies in order to keep

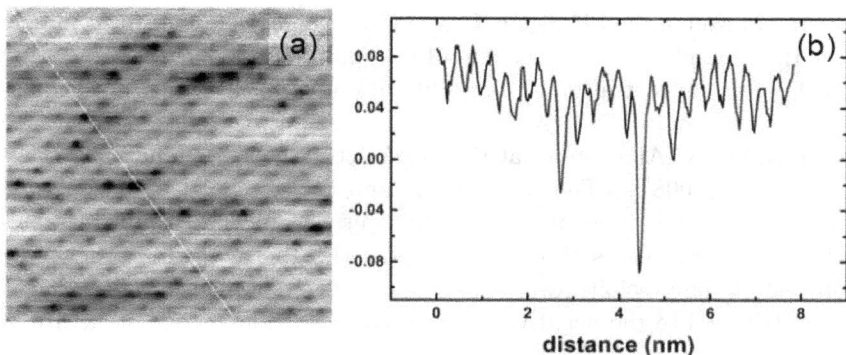

Fig. 4.2 (a) NC-AFM image of $KCl_{0.6}Br_{0.4}(001)$. (b) Cross-section corresponding to the white bar in the image. Reprinted from [Bennewitz *et al.* (2002)], Copyright 2002, with permission from Elsevier.

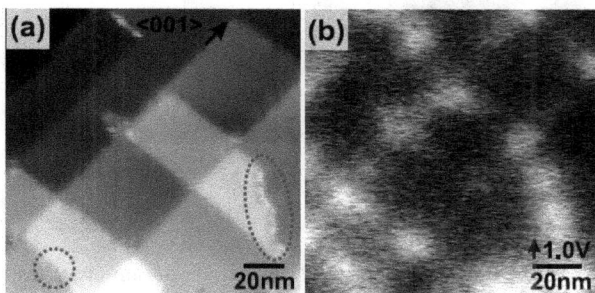

Fig. 4.3 (a) Topography and (b) potential images on a KCl(001) surface. Reprinted with permission from [Barth and Henry (2007)]. Copyright (2007) by the American Physical Society.

charge neutrality. As a consequence, a double layer is created, in which the negative net charge of the cation vacancies on the surface is compensated by the positive impurities below the surface, thus forming a surface dipole. The KPFM measurements clearly indicate that these negative cation vacancies are preferentially located at kink sites of steps, rather than on flat terraces.

By varying the impurity concentration, the atomic structure of ionic surfaces can be significantly reorganized. For instance, the precipitation of additional new phases may appear at certain critical concentrations, resulting in nanostructuring of the surfaces. An example of such phenomenon, characteristic of halide crystals, is the *Suzuki phase*, which was recognized in the 1950's for NaCl crystals with Cd^{2+} impurities [Suzuki (1955)]. This phase is characterized by a cubic structure twice larger than NaCl, high concentration of ordered cation vacancies, and inclusion of Suzuki precipitates in the NaCl matrix. A Suzuki phase has been recently observed by NC-AFM on a NaCl(001) surface with Mg^{2+} impurities [Barth and Henry (2008)]. The precipitates appeared embedded in the NaCl matrix, and each type of ions in the unit cell of the Suzuki structure could be identified, as shown in Fig. 4.4. Furthermore, a strong bright contrast at the precipitates was observed in corresponding KPFM images, and attributed to the negative cation vacancies in the Suzuki structure.

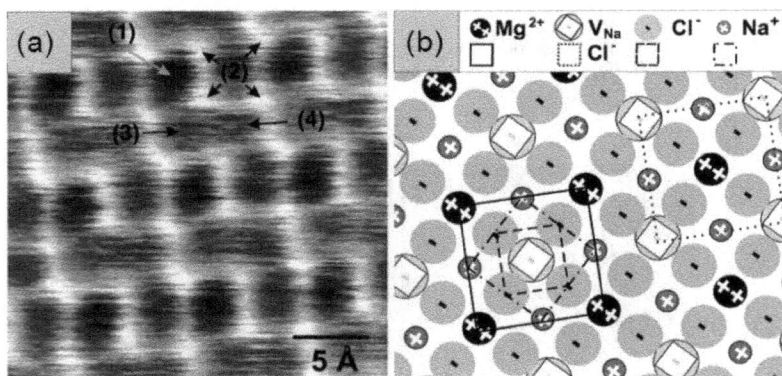

Fig. 4.4   (a) NC-AFM image (3.3 nm) with atomic resolution of a Suzuki precipitate. The dark spots (1) correspond to Mg^{2+} ions, the bright spots (2) to Cl^-. The sites (3) are occupied by Na^+ ions, and the vacancies are located in the middle of Mg^{2+} squares (4). (b) Schematic drawing illustrating the atomic structure of the precipitate. Reprinted with permission from [Barth and Henry (2008)]. Copyright (2008) by the American Physical Society.

4.1.2   *Alkaline earth halide surfaces*

The $CaF_2(111)$ surface was first imaged with atomic resolution in 1999 [Reichling and Barth (1999)]. On this surface, NC-AFM revealed a characteristic triangular pattern with apexes along the [121] and equivalent directions [Fig. 4.5(a)]. This pattern is caused by secondary maxima, which appear enhanced in the section in Fig. 4.5(b). Stable defects could not be observed till the surface was exposed to oxygen while scanning. After that, stable atomically resolved defects were recognized and associated to OH^- groups incorporated into the surface. A jump of one of the groups from

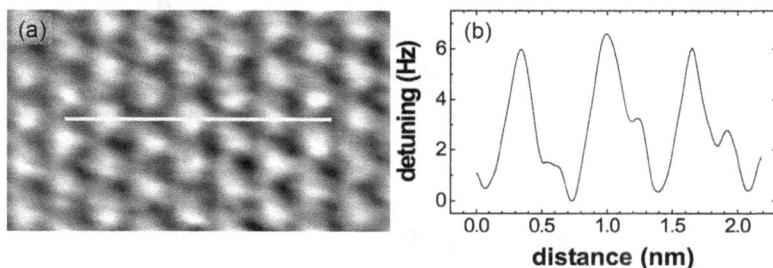

Fig. 4.5 (a) NC-AFM image $(3.0 \times 2.2 \text{ nm}^2)$ of a $CaF_2(111)$ surface acquired at constant tip height. (b) Cross section along the [121] direction, corresponding to the white line in the image. Reprinted with permission from [Foster *et al.* (2001)]. Copyright (2001) by the American Physical Society.

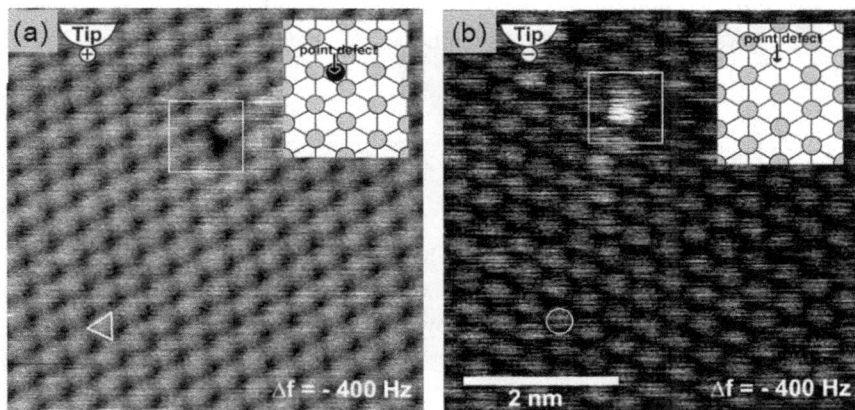

Fig. 4.6 Constant-height NC-AFM images of a point defect on a $CaF_2(111)$ surface. The insets illustrate the atomic contrast at the point defect. Analysis of sublattice contrast in (a) and (b) revealed that imaging in (a) and (b) was performed by a Ca^{2+}-terminated tip and a F^--terminated tip, respectively. From [Fujii and Fujihira (2007)]. Reproduced with permission from IOP Publishing.

one atomic cell to the next one was also noticed (see also Chapter 13). In another experiment a point defect was created by gently contacting a probing tip, which had been previously covered with CaF_2 by repeated indentations into the surface [Fujii and Fujihira (2007)]. The defect was either imaged as a dark depression or as a bright protrusion (Fig. 4.6). While imaging, reversible changes between the two types of contrast were observed, which was attributed to sudden exchanges in the terminating ion species at the tip apex. Atomic force microscopy investigations on other alkaline earth halide surfaces such as BaF_2 and $SrF_2(111)$ have been also reported [Barth and Reichling (2000)].

4.2 Oxide Surfaces

Although the AFM can in principle image true insulating oxides with atomic resolution, difficulties arising in the surface preparation and in the complexity of the surface structures have resulted in only few successful experiments, most noticeably on aluminum and magnesium oxides. This scenario changes for mixed conducting oxides, where the narrower band gap allows STM imaging after adequate thermal treatment of the samples. On these surfaces a large variety of structures and reconstructions have been reported, the interpretation of which is still subject of intense debate. According to the guidelines of the book, we will mainly focus on AFM results.

4.2.1 *True insulating oxide surfaces*

4.2.1.1 *Aluminum oxide*

We have mentioned in Sec. 1.2.1 that the α-$Al_2O_3(0001)$ surface may undergo a series of reconstructions. Atomically resolved images of one of these structures could be acquired by NC-AFM after heating the sample beyond 1200 °C [Barth and Reichling (2001)]. A grid of rhombic unit cells was observed, with a high degree of order at the center of each unit cell and disordered regions at the boundaries between unit cells (Fig. 4.7). A hexagonal arrangement of atoms was found at the center of each rhombus (not shown here). The authors also noticed that the disordered areas at the boundaries of the rhombi were preferential locations for chemical reactions when the surface was exposed to water and hydrogen (Fig. 4.8). Although the exact nature of these reactions could not be determined, it is plausible that the reactions corresponded to the onset of hydroxide formation.

4.2.1.2 *Magnesium oxide*

The MgO(001) surface is usually characterized by the presence of several extended defects and point defects. Extended defects are associated to steps, line defects due to missing rows of Mg^{2+} and O^{2-} ions, rectangular

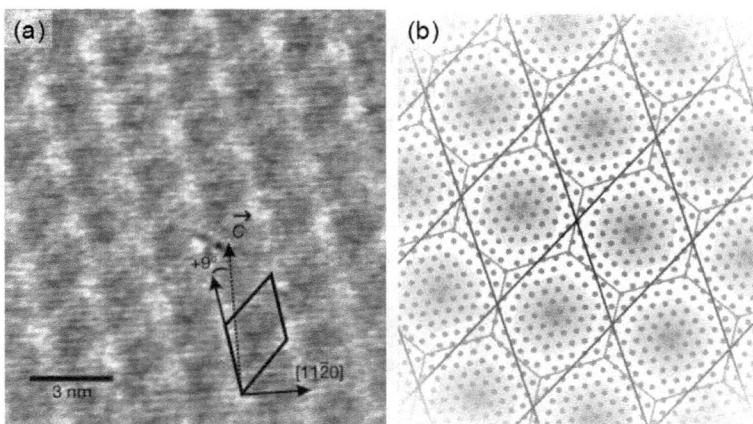

Fig. 4.7 (a) High resolution NC-AFM image of the high temperature $(\sqrt{31} \times \sqrt{31})R\pm9°$ reconstruction of an α-Al$_2$O$_3$(0001) surface. The rhombi represent the unit cell of the reconstructed surface and are tilted by 9° with respect to the c-axis of the alumina crystal. Stripes in the {1120} direction represent rows of Al atoms, which are the predominant species on this surface. Light regions between the rhomboid tiles represent disordered areas. (b) Superposition of the hexagonal domains formed by Al atoms with reconstruction rhombi. Reprinted by permission from Macmillan Publishers Ltd: Nature [Barth and Reichling (2001)]. Copyright 2001.

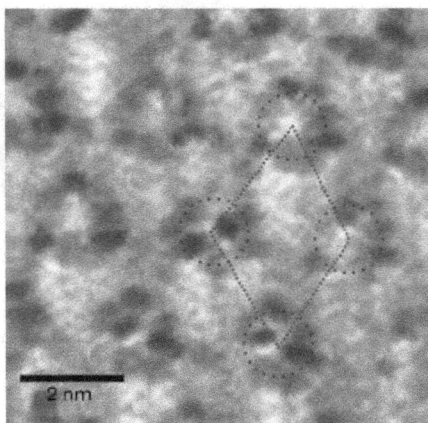

Fig. 4.8 The reconstructed α-Al$_2$O$_3$(0001) surface after exposure to 35×10^{-4} mbar·s of water and hydrogen originating from the residual gas of the UHV chamber. Hydroxide crystallizes into a pattern reproducing the rhomboidal structure of the reconstructed surface. A substructure in form of rings is also visible. Reprinted by permission from Macmillan Publishers Ltd: Nature [Barth and Reichling (2001)], copyright 2001.

holes of nanometer size produced in the cleavage process, and complex adstructures formed by adatoms. All these features could be identified by NC-AFM [Barth and Henry (2003); Ashworth *et al.* (2003)], as shown in Fig. 4.9. KPFM measurements on the adstructures also showed a strong contrast, due to different chemical composition and/or charging in the extended defects [Barth and Henry (2006b)]. Evidence of point defects was found in atomically resolved NC-AFM images (not shown here) and associated to low coordinated sites, vacancies, and impurity atoms.

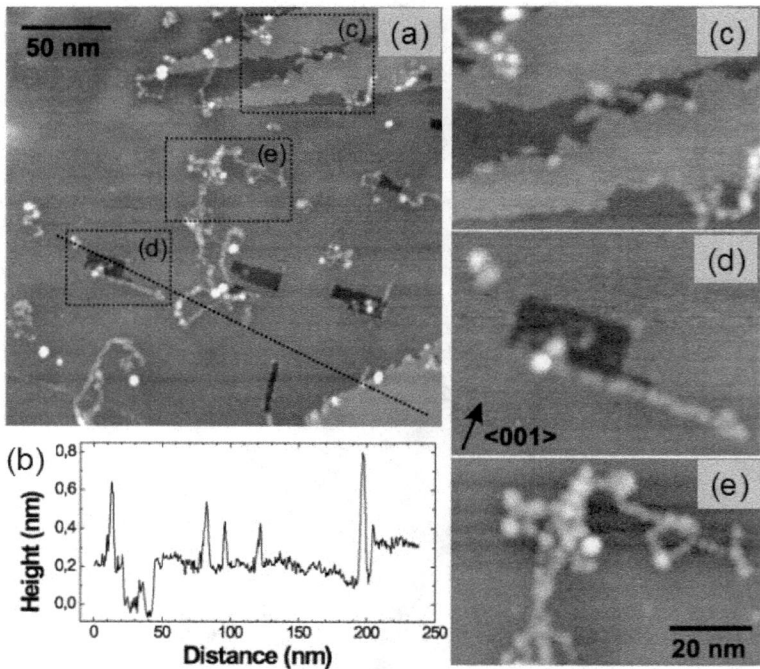

Fig. 4.9 (a) Large-scale NC-AFM image of a MgO(001) surface cleaved in UHV. (b) Cross-section corresponding to the dark dotted line in (a) showing the height of steps, holes, and adstructures. (c)-(e) Subimages of (a) showing major features of steps, holes and adstructures. Reprinted with permission from [Barth and Henry (2003)]. Copyright (2003) by the American Physical Society.

4.2.1.3 *Silicon dioxide*

Despite the importance of quartz surfaces, their complex structure remains poorly understood. Atomic force microscopy has been applied to investigate the formation of etch pits during dissolution [Gratz *et al.* (1991)] as well as morphological instabilities created during growth [Kawasaki (2003)], but only with low lateral resolution. A (1×1) reconstruction for α-quartz (0001) with 120° symmetry for the surface unit cell was suggested by theoretical considerations [de Leeuw *et al.* (1999); Rignanese *et al.* (2000)]. In this structure terraces with step heights of a multiple 1/3 of the unit cell height should lead to domains rotated by 60° relative to each other. A confirmation of these assumptions has recently come from AFM images (Fig. 4.10) on a quartz surface previously polished by diamond powder and treated by an elaborated procedure consisting of several cleaning and annealing stages [Steurer *et al.* (2007)].

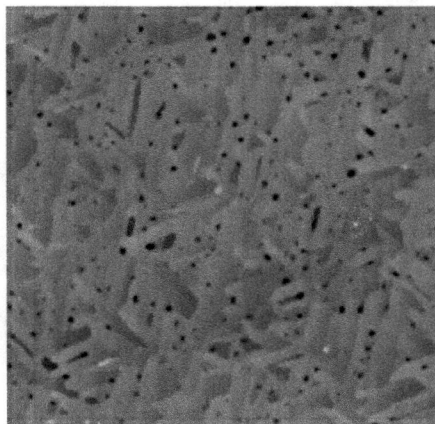

Fig. 4.10 AFM image (4.1 μm) of an α-quartz (0001) surface. The image shows large atomically flat terraces with an average width of about 200 nm and an average step height of about 1.9 nm. The terraces are oriented with angles of 60° relative to each other. Several pits with depths ranging between 2 and 5 nm can be also observed. The pits were presumably created by relaxation processes during the annealing. Reprinted from [Steurer *et al.* (2007)], Copyright 2007, with permission from Elsevier.

4.2.2 *Mixed conducting oxide surfaces*

4.2.2.1 *Titanium dioxide*

The TiO_2(110) surface has been repeatedly studied using STM and NC-AFM. Fig. 4.11 shows a typical image acquired by STM [Diebold *et al.* (1998)]. Bright and dark rows along the [001] direction are clearly visible. The distance between the rows corresponds to the unit cell size of 0.65 nm along the [1$\bar{1}$0] direction. After exposure to oxygen the bright spots in the dark rows disappeared. This indicates that the defects corresponded to oxygen vacancies, and, consequently, that the dark contrast in Fig. 4.11(a) is due to lines of bridging oxygen atoms while the bright contrast is produced between the rows at the Ti sites. Non-contact AFM images of a similar surface also exhibited a row structure with the dimensions of the ideal unit cell [Fig. 4.11(b)] [Fukui *et al.* (1997a)]. By applying a small bias on the tip, which canceled the potential difference between sample and tip, individual oxygen atoms could be resolved. Depending on the nature of the tip apex, different categories of NC-AFM images on $TiO_2(110)$ have been classified [Enevoldsen *et al.* (2007)].

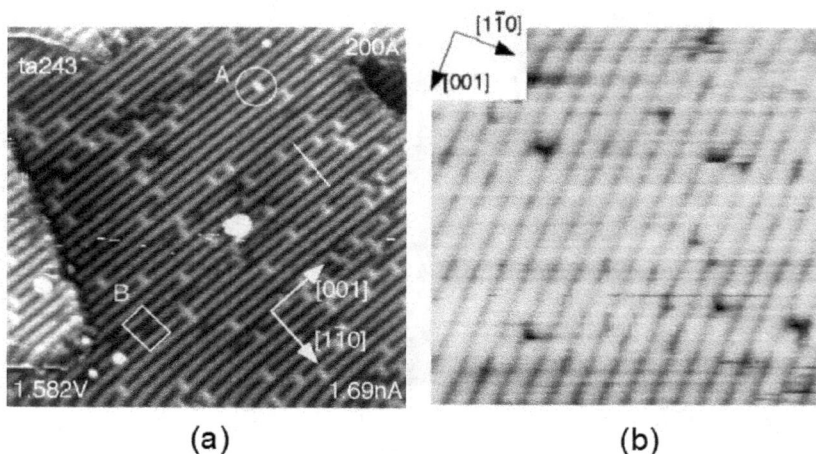

(a)   **(b)**

Fig. 4.11   (a) STM image (20 nm) of a $TiO_2(110)$-(1×1) surface. Bright spots (A) are vacancies in the bridging oxygens. Dark spots (B) are attributed to subsurface defects. Reprinted from [Diebold *et al.* (1998)], Copyright 1998, with permission from Elsevier. (b) NC-AFM image (10 nm) of a $TiO_2(110)$-(1×1) surface. Reprinted with permission from [Fukui *et al.* (1997a)]. Copyright (1997) by the American Physical Society.

The (1×2) reconstruction of the $TiO_2(110)$ surface is documented by the STM and NC-AFM images in Fig. 4.12. Rows with bright contrast run along the [001] direction with a constant separation of 1.3 nm, i.e. twice the row separation of the (1×1) substrate. These rows are linked with a cross-shaped structure (*cross-link*), which is aligned perpendicular to the rows [Fukui and Sakai (2006)].

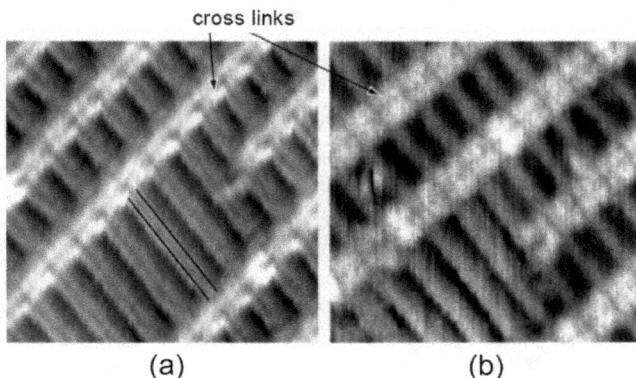

Fig. 4.12 (a) STM image (11 nm) and (b) NC-AFM image (12 nm) of a TiO$_2$(110)-(1×2) surface. Reprinted with permission from [Fukui and Sakai (2006)]. Copyright 2006 American Chemical Society.

4.2.2.2 *Zinc oxide*

The Zn-terminated face of ZnO has been also investigated by both STM and NC-AFM. STM studies suggested that this surface is stabilized by a reconstruction involving triangular structures (Fig. 4.13), but this reconstruction could not be resolved at the atomic scale [Dulub *et al.* (2003)]. Atomic resolution was indeed achieved by NC-AFM, giving strong evidence that the morphology of the resulting surface is significantly influenced by the preparation temperature [Ostendorf *et al.* (2008)]. In fact, while triangular reconstructions without any apparent ordering among themselves stabilized surfaces annealed at 875 °C, a complex combination of faceting and triangular reconstruction appeared upon annealing beyond 1025 °C.

4.2.2.3 *Tin dioxide*

Stoichiometric SnO$_2$ surfaces are particularly problematic, since they can only be formed at elevated oxygen pressures or using efficient oxidants such as ozone, nitrogen, or oxygen plasma. As a consequence, surface science experiments on these surfaces under UHV are not so conclusive. A surface unit cell consistent with a bulk truncation was recognized in STM images on a SnO$_2$(101) surface prepared by sputtering and annealing at 625 °C [Batzill *et al.* (2004)]. Other techniques suggested that this surface has the reduced structure shown in Fig. 1.6(b).

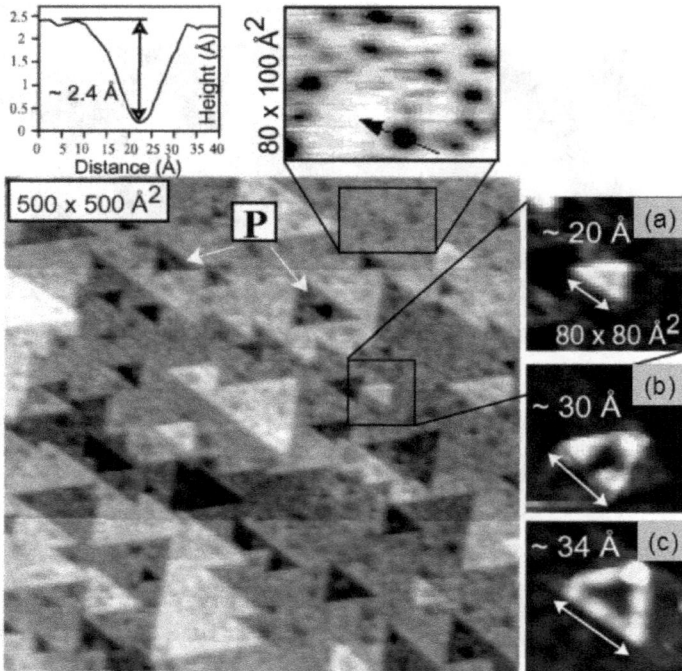

Fig. 4.13   STM images of the Zn-terminated (0001) surface of zinc oxide. Terraces are covered by triangular islands and holes that are separated by single layer height steps. A high concentration of small holes is present on the terraces (see inset). The images (a)-(b) show small islands with shapes that are characteristic for each island's size. Reprinted with permission from [Dulub *et al.* (2003)]. Copyright (2003) by the American Physical Society.

4.2.2.4   *Cerium dioxide*

Early STM and NC-AFM studies of CeO_2 at room temperature revealed various oxygen vacancy defects, but on a small scale and with limited resolution, which precluded a discussion of their relative distribution [Nörenberg and Briggs (1997); Fukui *et al.* (2002)]. More recently, an STM investigation on a $CeO_2(111)$ surface at high temperature established that single subsurface oxygen vacancies can nucleate linear surface oxygen vacancy clusters (Fig. 4.14). This important observation can explain the catalytic properties of CeO_2, since oxygen release is clearly enhanced by this process. Surface and subsurface oxygen vacancies on $CeO_2(111)$ have been also identified by further NC-AFM investigations at 80 K [Torbrügge *et al.*

(2007)]. By combining two complementary signals, namely the topography and the damping, oxygen vacancies buried at the third surface atomic layer could be also located.

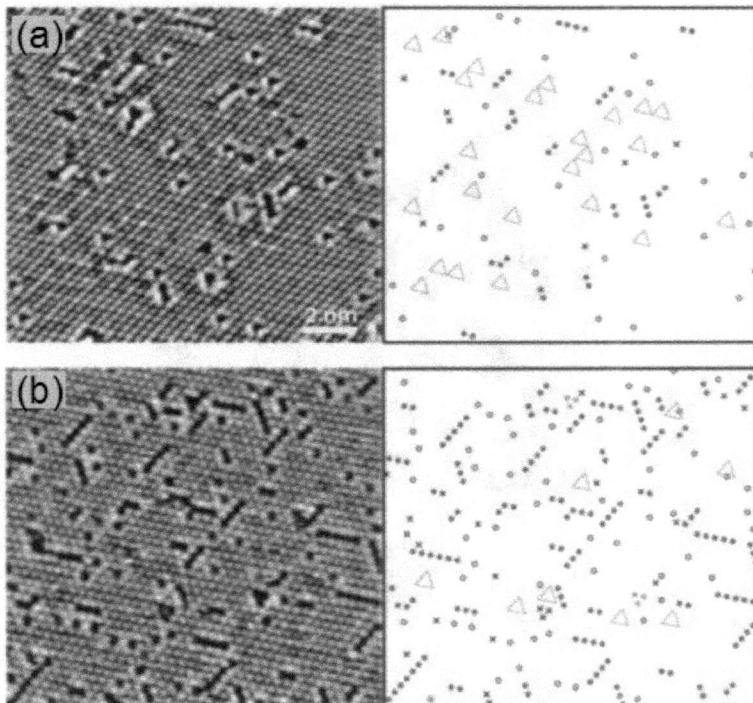

Fig. 4.14 STM images of a $CeO_2(111)$ surface obtained (a) after 1 min and (b) after 5 min of annealing at 900 °C, with corresponding representations of the observed defects. From [Esch *et al.* (2005)]. Reprinted with permission of AAAS.

4.2.2.5 *Strontium titanate*

Many reconstructions of the $SrTiO_3(001)$ surface have been experimentally observed. Also in this case the structures of the reconstructed surfaces are strongly influenced by the annealing conditions, e.g. atmosphere (oxidation or reduction), temperature and time, as well as by the surface conditions before annealing. One of the first stable structures, obtained after annealing in ambient and in UHV conditions, consisted of steps with height equal to integer multiples of the unit cell and a row-like structure

(Fig. 4.15) [Jiang and Zegenhagen (1999)]. This structure remained stable after exposure to oxygen or ambient. Other reconstructions resulted from annealing of surfaces that had been previously etched by hydrofluoric acid in UHV [Castell (2002)]. Ordered structures caused by the adsorption of Sr adsorbed on a TiO_2 surface are shown in Fig. 4.16 [Kubo and Nozoye (2001)].

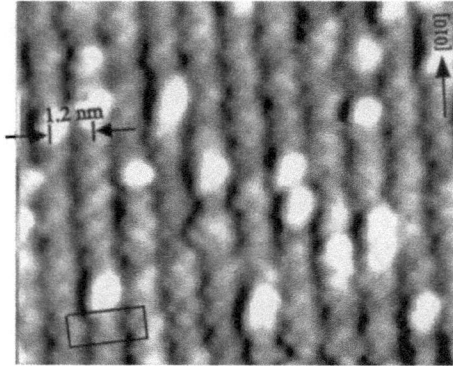

Fig. 4.15   STM image of the $c(6 \times 2)$ reconstruction of $SrTiO_3(001)$. The spacing between adjacent rows is 1.2 nm. Some features within rows are locally resolved. A unit cell of the $c(6 \times 2)$ is outlined in the lower left corner of the image. Reprinted from [Jiang and Zegenhagen (1999)], Copyright 1999, with permission from Elsevier.

4.3   Modeling AFM on Bulk Insulating Surfaces

4.3.1   *Halide surfaces*

The model introduced in Sec. 3.3 has been successfully used to reproduce NC-AFM images of $CaF_2(111)$ [Foster *et al.* (2001, 2002)]. The model tip consisted in a conducting cone with an oxide nanotip at the spherical cap terminating the cone. A theoretical image of the CaF_2 surface simulated with a nanotip terminated by a Mg^{2+} ion is shown in Fig. 4.17. The smallest tip-surface separation was chosen to match the average frequency shift used in the experimental image in Fig. 4.5. The maximum frequency shift corresponds to the most protruding F^- ions in the surface, $F(2)$. At such locations the interaction is dominated by the strong attraction between the positive potential from the tip and the negative potential over the F^- ions. The contrast is further increased by the displacement of the F^- ions

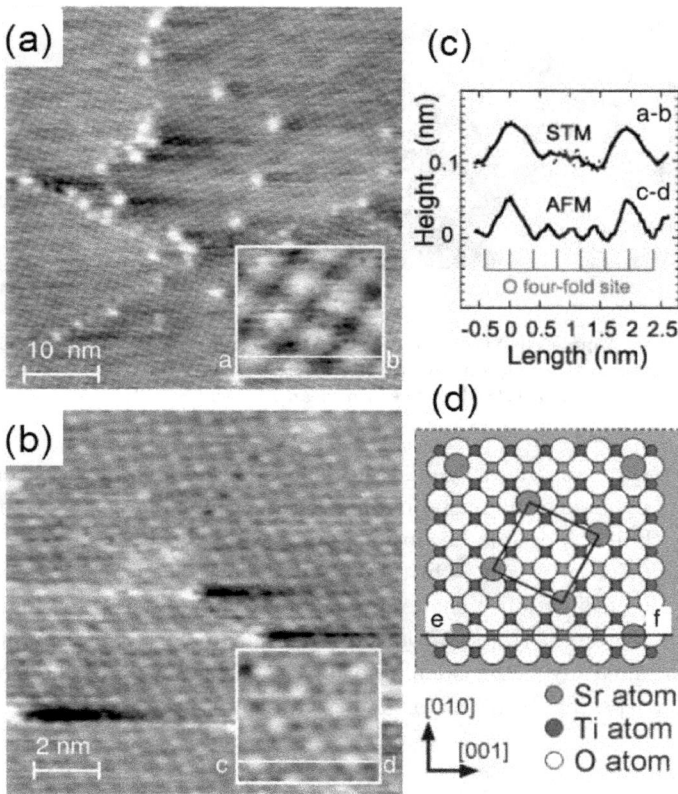

Fig. 4.16 (a) STM and (b) NC-AFM images of $SrTiO_3(001)$ after heating at 1200 °C for several seconds. (c) Cross-sections along the lines a-b and c-d. (d) Proposed model of the $SrTiO_3(001)$-$(\sqrt{5} \times \sqrt{5})$-R26.6° surface reconstruction. The line e-f corresponds to the lines a-b and c-d, respectively. Reprinted with permission from [Kubo and Nozoye (2001)]. Copyright (2001) by the American Physical Society.

due to interaction with the tip. The interaction with the deeper F^- ion, $F(3)$, which could be also distinguished, is responsible for the triangular shapes observed in the experiments, with slower contrast decay compared to the contrast change between $F(2)$ and $Ca(1)$ sites. Simulations with an O^{2-} ion at the apex of the nanotip were also performed in order to model a tip with a negative electrostatic potential due to an oxide layer or contamination by an anion. As a result, spots with disk-like shapes were observed, which do not have a counterpart in the experiments.

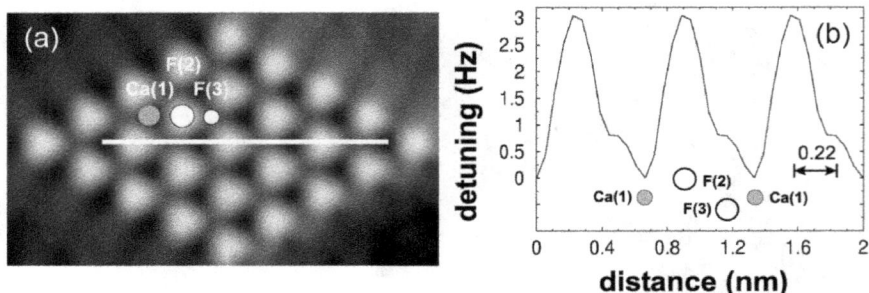

Fig. 4.17   (a) Simulated NC-AFM image (3.0×2.2 nm^2) and (b) cross-section of a CaF$_2$(111) surface taken at constant tip height. Reprinted with permission from [Foster *et al.* (2001)]. Copyright (2001) by the American Physical Society.

The results of these simulations show the sign of the potential on the CaF$_2$(111) surface can be deduced from the asymmetry of the image contrast. This is possible because the unit cell contains one Ca^{2+} ion and two inequivalent F$^-$ ions. On the (001) surfaces of alkali halide surfaces, sublattice identification is made difficult by the symmetric arrangement of the ions. However, the two ion sublattices can be still distinguished by comparing measured and calculated forces as a function of the tip-sample distance at specific locations, as shown in Chapter 11.

4.3.2   *Oxide surfaces*

The interaction between Al$_2$O$_3$, TiO$_2$ and MgO surfaces and Si dangling-bond tips has been studied using *density functional theory* (DFT) [Foster *et al.* (2004)]. The strongest interaction was observed at the highest anions in the surfaces, where chemical bonds tend to be formed. The model MgO nanotip previously discussed has been also used to simulate NC-AFM imaging on a MgO(001) surface [Livshits *et al.* (1999)] and on a TiO$_2$(110) surface with various defects [Lauritsen *et al.* (2006)]. Also in the case of oxide surfaces the image contrast was found to depend crucially on the tip-termination polarity. As shown in Fig. 4.18, different atomic arrangements at the tip apex produce two types of atomically resolved images. However, neither of them reveals the true topography of the surface, or its defects and adsorbates.

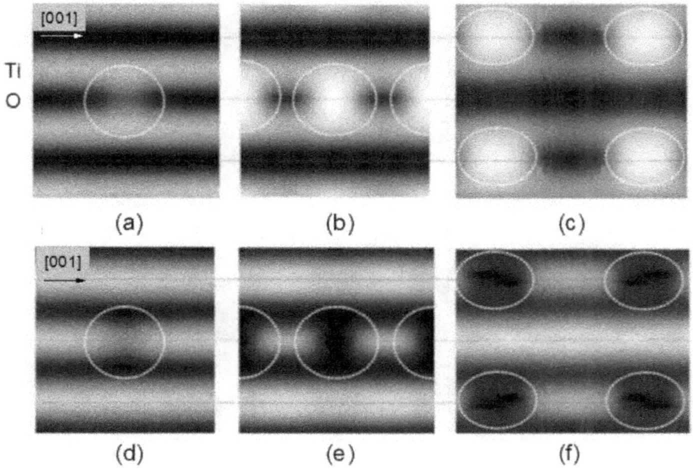

Fig. 4.18   Simulated NC-AFM images of a $TiO_2(110)$ surface. The upper and lower rows correspond to negative and positive terminated tips. Images (a,d) show single oxygen vacancies, (b,e) single hydroxyls (OH), and (c,f) double hydroxyls (pairs of OH) on the surface. From [Lauritsen *et al.* (2006)]. Reproduced with permission from IOP Publishing.

Chapter 5

Scanning Probe Microscopy on Thin Insulating Films

Ultrathin films of insulating materials on a single crystal often exhibit unusual properties compared to the corresponding bulk-truncated surfaces. Depending on the interaction with the substrate, the adsorbates can form either flat layers or three dimensional structures. Regular nanopatterns are observed when ionic films are deposited on vicinal surfaces or on other ionic films with different chemical composition. Insulating films grown on a conducting surface can be imaged not only by AFM but also by STM, which gives important information on the electronic structure of these systems.

5.1 Halide Films on Metals

5.1.1 *Carpet-like growth*

The first atomically resolved images of ultrathin films of alkali halides grown on metals were recorded by STM on a NaCl film deposited on an Al(111) surface at room temperature [Hebenstreit *et al.* (1999)]. At low coverages flat NaCl islands were formed, where the NaCl dipoles seemed to lay flat. A 'carpet-like' growth across the surface steps can be recognized in Fig. 5.1(a). The rectangular shape of the islands reveals a strong tendency to minimize the number of corner sites, and the mobility of the NaCl dipoles on the surface. The formation of a second and a third layer started before completion of the first layer at a total coverage of approximately 0.4 ML (Volmer-Weber growth, see Sec. 2.3). Atomically resolved images with a periodicity of 0.4 nm [Fig. 5.1(b)] showed that STM can distinguish only one species of ions, as NC-AFM on bulk-truncated surfaces (Sec. 4.1). In the present case, complementary theoretical calculations suggested that the highest density of states around the Fermi edge corresponds to the locations of Cl^- ions, so that that this species should appear as protrusions in the STM images. The reported atomic corrugation, in the range of few tens of pm, was always larger on top of bilayer islands than on a monolayer, independently of the tip conditions [Fig. 5.1(c)]. At a thickness of three layers or more, no atomic resolution could be achieved. Furthermore, atomic resolution could not be simultaneously obtained on film and substrate, since the Al(111) surface required a much lower tunneling voltage to be imaged.

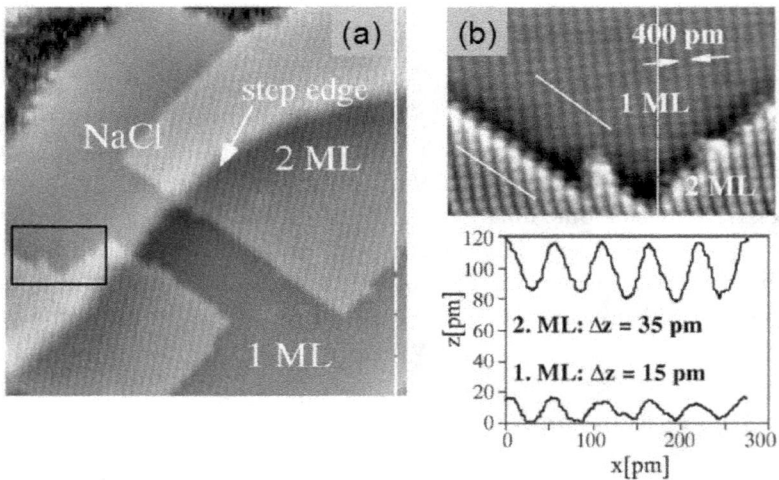

Fig. 5.1   (a) STM image (49 nm) of NaCl islands grown across a step edge of the Al(111) substrate. (b) Atomically resolved part of (a) showing a single and a double layer (10 nm). The atomic corrugation appears larger on the double layer. Reprinted from [Hebenstreit *et al.* (1999)], Copyright 1999, with permission from Elsevier.

Non-contact AFM has been applied to investigate ultrathin films of NaCl on Cu(111) [Bennewitz *et al.* (1999, 2000)]. Also in this case, NaCl was found to form rectangular islands, and monatomic steps of the substrate were covered by a continuous film in a carpet-like fashion. Atomically resolved images of a NaCl island are shown in Fig. 5.2. Contrast is enhanced at the edges of the island compared to the terrace, and the maximum contrast is observed at kink sites. In a damping image acquired at the same time the atomic contrast at step edges is even more pronounced [Fig. 5.2(b)]. These results are clearly due to the lower coordination at edge and kink sites, although a clear interpretation of the damping mechanism, which strongly depends on the tip conditions, is difficult. A comparison between Fig. 5.2(a) and (b) in fact reveals how a sudden tip change affected much more the damping signal than the corresponding topography. A strong contrast between substrate and film is also visible, which is clearly due to the difference in the work function between the two materials. Whether the NaCl dipoles were adsorbed flat on the surface, or perpendicular to it, could not be revealed by NC-AFM images alone.

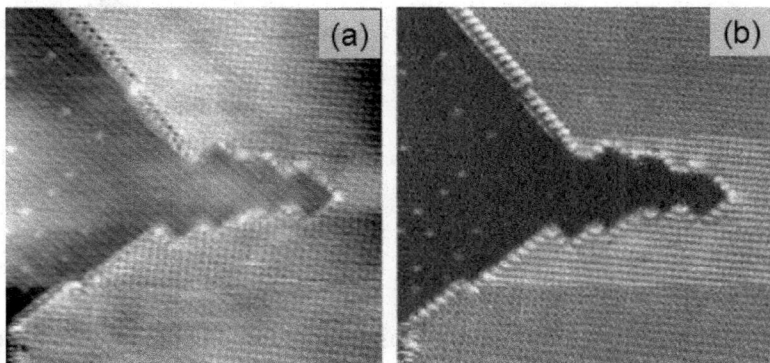

Fig. 5.2   (a) NC-AFM topography and (b) corresponding damping image of a NaCl island on a Cu(111) surface (18 nm). The tip changed after one-fourth of the scan, thereby changing the contrast in topography and increasing the damping contrast. After two-thirds of the scan, the contrast from the lower part of the images was reproduced, indicating that a reversible tip change. The bright dots on the copper substrate were attributed to single sulfur impurity atoms. Reprinted with permission from [Bennewitz *et al.* (2000)]. Copyright (2001) by the American Physical Society.

The growth of alkali halides on metal surfaces may modify the electronic structure of the substrate and produce characteristic *Moiré patterns*. These patterns are caused by displacements of the charged ions, which modulate the electrostatic potential seen by the surface electrons, and can be revealed by STM. Fig. 5.3 shows a Moiré pattern formed by NaCl islands grown on Ag(001) [Pivetta *et al.* (2005)]. The islands present two different orientations, and the pattern is visible only on islands tilted by 45° with respect to the main crystallographic directions of the substrate. Another example of Moiré pattern was reported on the NaCl/Cu(111) system [Repp *et al.* (2004b)].

Alkali halide thin films on metals were also investigated by KPFM [Loppacher *et al.* (2004)]. After depositing four different alkali chlorides on the Au(111) surface, the changes in the local work functions, ΔV_{cpd}

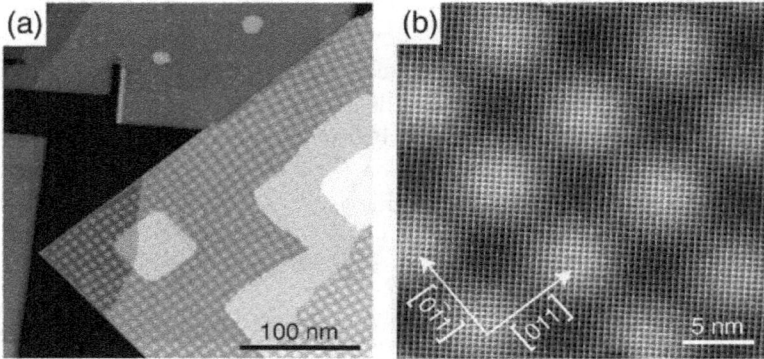

Fig. 5.3   (a) STM image (300 nm) showing NaCl islands grown on Ag(001) with 0°
and 45° orientations. (b) STM image (25 nm) acquired on a 45° island with atomic
resolution of the NaCl adlayer. Reprinted with permission from [Pivetta *et al.* (2005)].
Copyright (2001) by the American Physical Society.

(referred to the value on the bare substrate) were compared and a striking
linear dependence between ΔV_{cpd} and the cationic radii of the alkali species
was found (Fig. 5.4).

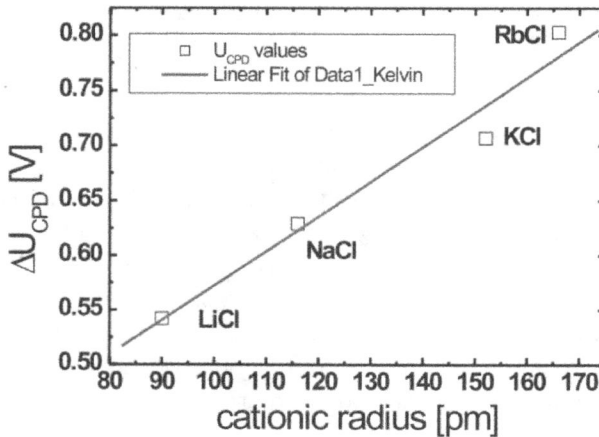

Fig. 5.4   Experimental values of ΔV_{cpd} for LiCl, KCl, NaCl and RbCl thin films on
Au(111) versus the ionic radii of the corresponding cations. From [Loppacher *et al.*
(2004)]. Reproduced with permission from IOP Publishing.

5.1.2 *Restructuring and patterning of vicinal surfaces*

Vicinal surfaces are stepped surface formed when a crystal is deliberately miscutted. These surfaces are subject to reconstructions, which produce wide terraces separated by monatomic steps. The growth of NaCl on a vicinal Cu(211) surface, formed by (111) terraces and intrinsic (100) steps, was investigated by STM at room temperature and growth temperature of 95 °C [Fölsch *et al.* (2000, 2002a)]. The film deposition transformed the initially flat surface into a periodic one-dimensional hill-and-valley structure consisting of (311) and (111) facets. Sodium chloride was found to grow selectively only on the (311) facets, creating a regular surface pattern with alternating stripes of bare Cu and areas covered by NaCl (Fig. 5.5). The spacing between adjacent stripes could be tuned in a range of several nanometers by varying the growth and annealing temperature.

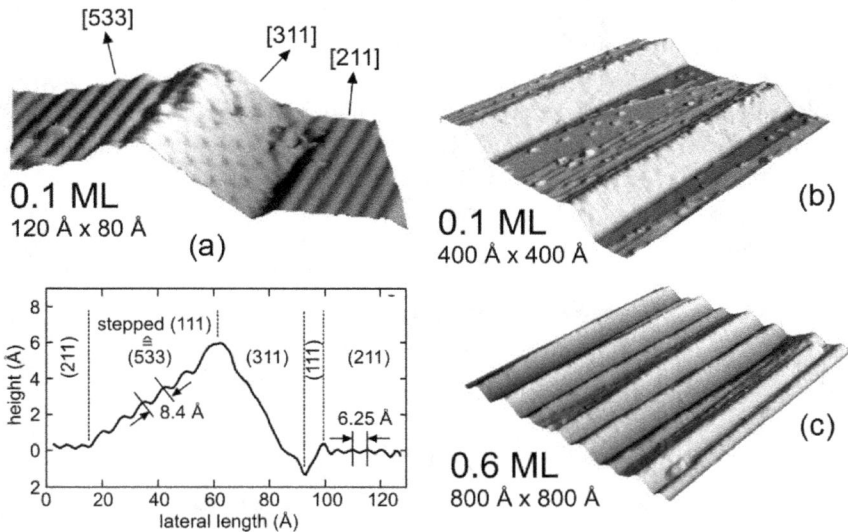

Fig. 5.5 (a) STM image of a Cu(211) surface after deposition of 0.1 NaCl monolayers. A structure of (311) and (111) facets started to be formed, with only the (311) facets covered by NaCl. The stepped (111) facet is (533)-oriented at this initial stage. The lower panel of (a) shows a cross-section perpendicular to the intrinsic step direction. (b) Enlarged area showing two NaCl-covered (311) facets uniform in width and formed along the intrinsic step direction. (c) STM image after completion of the faceting process at 0.6 ML coverage showing an anisotropic and regular surface topography. Reprinted from [Fölsch *et al.* (2002a)], Copyright 2002, with permission from Elsevier.

The direct growth of NaCl on a Cu(311) surface was also investigated at growth temperatures above 130 °C [Repp *et al.* (2001)]. Under such conditions the second layer of NaCl appeared only after completion of the first layer, which clearly indicates a strong interaction between substrate and adsorbate. At higher coverages the growth slightly deviated from perfect layer-by-layer growth, since the third and fourth layer started to grow before the second layer had been completed. The shape of the islands in the first layer was determined by defect steps of the substrate. Perpendicular to the intrinsic steps the lattice constant of the NaCl adlayer adapted to that of the substrate, leading to a significant stretching of more than 6% with respect to the bulk value. The STM images also showed that the Cl^- ions were located on top of the intrinsic step edges of the Cu(311) surface, which was also suggested more recently by first principle calculations [Olsson *et al.* (2005)].

The previous results can be explained by a *charge modulation* process occurring on the stepped surfaces. Positive charges are expected along the intrinsic steps of the highly corrugated Cu(311) surface, while negative charges should be located between these steps. Thus, columns of Cl^- (or Na^+) ions can align with stripes of opposite charge at steps (or troughs), resulting in a strong electrostatic interaction between the NaCl adlayer and the substrate (Fig. 5.6). This interaction is strong enough to stabilize a considerable stretching perpendicular to the Cu rows. The same effect also explains the driving force for the NaCl-induced reorganization of the Cu(211) surface.

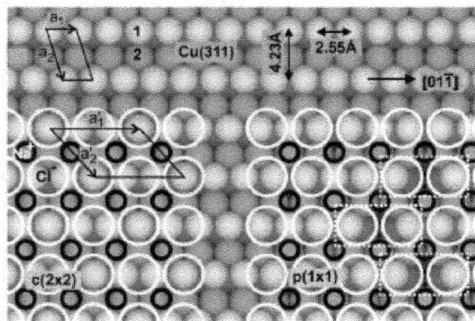

Fig. 5.6 Sphere model illustrating the Cu(311) surface with alternating monatomic (111) and (100) nanofacets. The (100)-terminated NaCl adlayer is represented by white circles (Cl^-) and black circles (Na^+). Reprinted with permission from [Repp *et al.* (2001)]. Copyright (2001) by the American Physical Society.

When NaCl was deposited on a different substrate — a kinked Cu(532) surface — regular assemblies of three-sided pyramids were observed (Fig. 5.7). However, only two pyramid faces appeared covered by NaCl. The chemical selectivity of this surface was tested by exposing the sample to carbon monoxide. As a result, adsorption took only place on the bare surface [Cu(111)], where a characteristic (4×4) superstructure, already known in the literature [Bartels *et al.* (1999)], was formed.

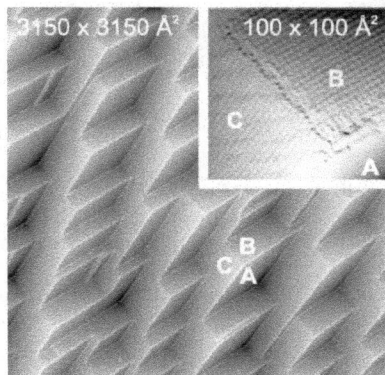

Fig. 5.7   STM image of a reorganized Cu(532) surface after growing 0.6 NaCl monolayers at 325 °C showing three well defined facet orientations. The inset shows that the facets B and C were covered by (100)-terminated NaCl, while the facet A was still covered by bare copper. Reprinted with permission from [Fölsch *et al.* (2002b)]. Copyright (2002) by the American Physical Society.

5.1.3   *Fractal growth at low temperatures*

The growth of LiF on Ag(111) at low temperature (77 K) was investigated by STM [Braun *et al.* (2000); Farias *et al.* (2000)]. A typical image recorded on a single terrace is shown in Fig. 5.8(a). Here, several *fractal islands* with a mean diameter of 25 nm can be recognized. The apparent height of the islands is significantly lower than the bulk distance between Li^+ and F^-, suggesting that the LiF dipoles lay flat on the surface. Furthermore, almost all fractals show a second layer grown around the nucleus. These results differ quite significantly from similar investigations on metal fractals [Brune (1998)]. The alkali halide fractals have also no symmetry, neither the threefold symmetry of the substrate nor the cubic symmetry of the bulk-truncated LiF(001) surface.

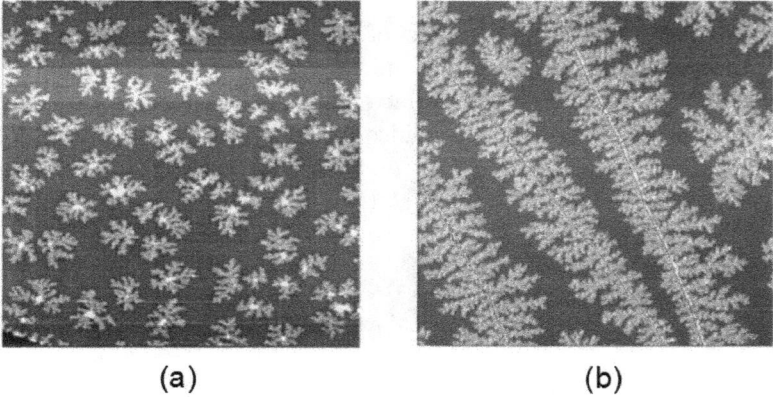

Fig. 5.8 (a) STM image (240 nm) revealing the fractal character of LiF islands formed on Ag(111) at 77 K. Reprinted from [Braun *et al.* (2000)], Copyright 2000, with permission from Elsevier. (b) STM image (300 nm) across a step edge. The same gray scale is used for all terraces. The highest terrace is the one on the top-right corner. Reprinted from [Farias *et al.* (2000)], Copyright 2000, with permission from Elsevier.

The island density n is related to the deposition flux F and to the diffusion coefficient $D = D_0 \exp(-E_m/k_B T)$, where E_m is the migration barrier of LiF on Ag(111), by the following scaling law [Venables *et al.* (1984)]:

$$n = \eta(\theta) \left(\frac{D}{F} \right)^{-1/3}. \tag{5.1}$$

In equation (5.1), the dimensionless factor η depends on the coverage θ. Assuming that the smallest stable cluster is a dimer, this quantity takes a value of 0.25 in the conditions of the experiment. According to the Einstein's relation, the prefactor $D_0 = \nu_0/2d$, where d is the dimension of motion and ν_0 is the attempt frequency for surface diffusion. Assuming a typical value for the attempt frequency $\nu_0 \sim 10^{11}\text{-}10^{12}$ s^{-1}, the migration barrier for diffusion of LiF monomers on Ag(111) terraces can be estimated in the range of several tens of meV. The fractal dimension, deduced from a statistical analysis on the mass-length relation of several aggregates, is $d = 1.75$. As a result, the observed island density distribution at terraces is consistent with the nucleation theory for a critical cluster size $i = 1$. This implies that LiF dimers are the smallest stable clusters at 77 K, and that LiF monomers are the diffusing units giving rise to the structures experimentally observed.

The nucleation of LiF molecules occurred on both the upper and the lower side of step edges for a large range of incidence fluxes [Fig. 5.8(b)]. This behavior is also quite different from metal-on-metal growth, where nucleation is restricted to the lower side of step edges due to the energy gain experienced by metal atoms when adsorbed at high-coordinated sites [Norskov (1990)]. Furthermore, comparable coverages were observed at both sides of the step edges. The fact that the two sides acted as equivalent nucleation centers for the molecules implies that a strong electric field was formed parallel to the surface and peaked at the step edges.

5.2 Halide Films on Semiconductors

As mentioned in Sec. 1.1, the lattice misfit between Si and CaF_2 is quite small. For this reason, CaF_2 was chosen as first candidate in earlier STM studies on insulating films grown on a semiconductor. Atomic resolution on $CaF_2/Si(111)$ structures could be already achieved by STM in 1989 [Avouris and Wolkow (1989)], whereas atomically resolved AFM images of a similar system were obtained much later [Klust *et al.* (2004)]. These images (not shown here) did not differ significantly from those acquired on bulk-truncated $CaF_2(111)$ surfaces with a positively terminated tip (Sec. 4.1). In another study, a stepped substrate was used to produce striped phases of CaF and CaF_2 with a periodicity of about 15 nm [Viernow *et al.* (1999b)]. More precisely, a 1.1° vicinal miscut of a Si(111) surface was covered by a monolayer of CaF, which formed a strong chemical bond with the silicon substrate. The next half monolayer of CaF_2 formed stripes on top of the CaF layer along the underlying steps (Fig. 5.9). The difference in the electronic states of the two materials, namely a lower conduction band minimum for the CaF stripes, could be determined by locally resolved tunneling spectroscopy (see Chapter 11).

STM studies on alkali halide films formed on semiconductor surfaces have been also reported. The growth of NaCl on Ge(001) at 150 K did not result in any preferential nucleation of islands at monatomic steps [Glöckler *et al.* (1996)] [Fig. 5.10(a)]. Some islands were found across step edges, indicating that the internal cohesion of the film was much stronger than the interaction with the substrate. From the apparent atomic corrugation measured by STM, the authors estimated that the first layer was a double layer, with the NaCl molecules adsorbed perpendicular to the substrate in an alternate fashion. The nucleation of a second NaCl layer before

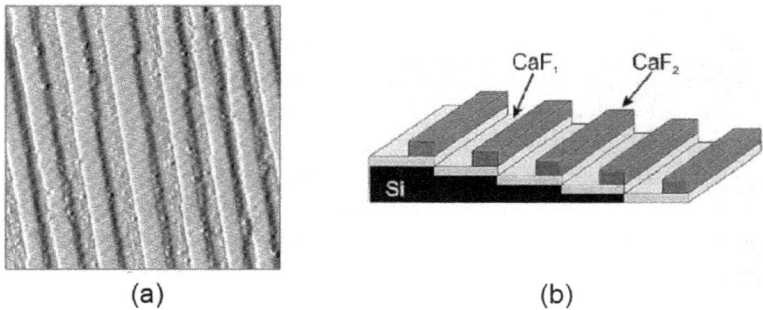

Fig. 5.9 (a) STM image (100 nm) of a Si(111) surface patterned with CaF_2/CaF stripes. Reused with permission from [Viernow *et al.* (1999b)]. Copyright 1999, American Institute of Physics. (b) Schematic model of the structure. Reprinted from [Rauscher *et al.* (1999)], Copyright 1999, with permission from Elsevier.

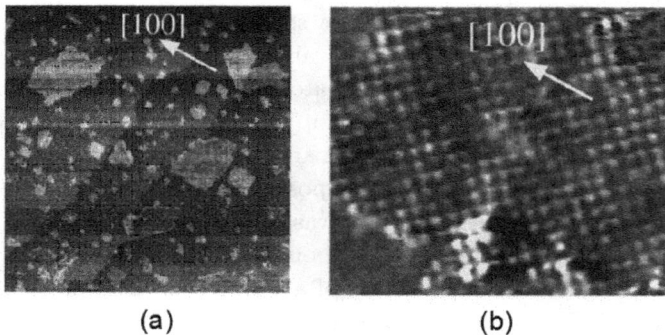

Fig. 5.10 STM images of NaCl on a Ge(001) surface (a) on large scale (113 nm) and (b) at atomic scale (10 nm). Reprinted with permission from [Glöckler *et al.* (1996)]. Copyright (2001) by the American Physical Society.

completion of the first double layer could be also observed, with the NaCl molecules assembled in chainlike ensembles. Atomically resolved images on the initial double layer showed once again that only one type of ions can be resolved by STM [Fig. 5.10(b)].

Another STM investigation addressed the growth of thin films of KBr on InSb(001) at 95 °C [Kolodziej *et al.* (2002)]. A complementary NC-AFM analysis allowed to extend the coverage over 100 monolayers. At very low coverages, islands of monatomic thickness were observed. These islands were often cut along the ⟨110⟩ crystallographic direction and their distribution on the substrate did not reveal specific orientations, which indicates that KBr dipoles diffused anisotropically on the surface. After completion of the first monolayer, the material in excess formed rectangular islands with edges oriented along the main crystallographic directions. At higher coverages the film was basically found to grow in a layer-by-layer fashion but, due to slow diffusion of KBr dipoles down across steps, the $(n+1)$-th layer started to grow before the completion of the n-th layer. As a result, pyramidal structures with rectangular bases were formed (Fig. 5.11). After thermal annealing, these rough structures were transformed into flat films exposing large (001) terraces. KBr islands could be unequivocally identified in potential images acquired on another KBr film grown on InSb(001) with submonolayer coverage, highlighting the chemical capabilities of the KPFM technique [Krok *et al.* (2004b)].

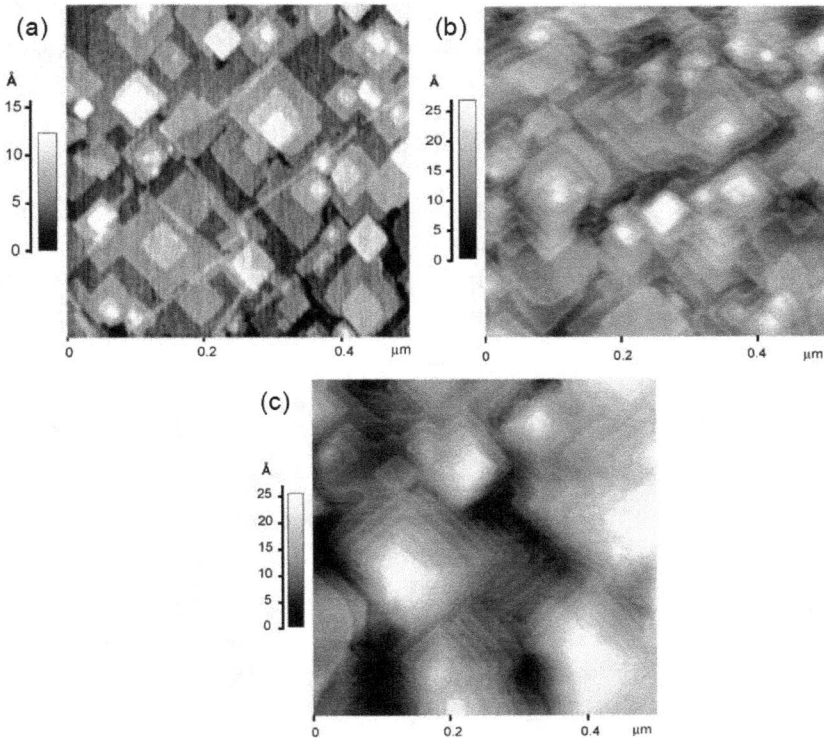

Fig. 5.11   NC-AFM images (0.5 μm) of several KBr/InSb(001) films with different average thickness: (a) 10 ML, (b) 60 ML, (c) 120 ML. At high coverages, deposited KBr tends to form regular pyramidal structures. Reprinted from [Kolodziej *et al.* (2002)], Copyright 2002, with permission from Elsevier.

5.3   Heteroepitaxial Growth of Alkali Halide Films

The large mismatches among lattice constants in alkali halide crystals make the heteroepitaxial growth of thin films of these materials quite intriguing. Here we discuss the case of KBr on NaCl(001). With a lattice mismatch of 17% the ratio of nearest-neighbor distances of the two crystals is very close to 7:6. The equilibrium structure of ultrathin KBr films was addressed by Monte Carlo simulations for film thickness varying between 2 and 6 monolayers [Baker and Lindgård (1996)]. Due to the large misfit, harmonic potential approximations could not be applied, and different interactions had to be taken into account at the same time. As a result, the film was found to 'rumple' with a periodicity of 6 KBr unit cells, as shown in Fig. 5.12. It is noticeable that the top layers of NaCl presented larger corrugation than the KBr films, despite of the fact that NaCl is the stiffer material in the bulk form. Furthermore, the first KBr layer appeared less corrugated than the layers over it. The rumpling effect has been further confirmed by NC-AFM measurements [Maier *et al.* (2007)]. In the topography images the corrugation of the KBr superstructure was too small, compared to the island height, to become visible, but it could be clearly revealed in the damping images recorded simultaneously (Fig. 5.13). Contact AFM images on the same system are described in Chapter 12, where the influence of the rumpling effect on friction is discussed. The growth of the 'reciprocal' system, i.e. NaCl on KBr(001), was also

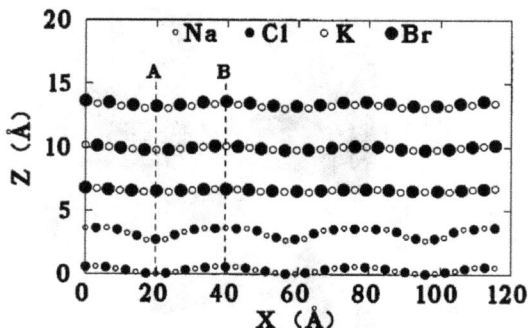

Fig. 5.12   Monte Carlo simulations of a $\langle 100 \rangle$ slice of a structure formed by the relaxation of 3 KBr monolayers on a NaCl(001) substrate at 100 K. The dashed lines A and B correspond to a minimum and a maximum of the corrugation, respectively. Reprinted with permission from [Baker and Lindgård (1996)]. Copyright (1996) by the American Physical Society.

Fig. 5.13 (a-b) NC-AFM topography images (200 nm) of a NaCl(001) surface partially covered with KBr. The typical size of the KBr islands with two and three layers height, respectively, depends on the evaporation rate: (a) corresponds to a slower rate of 0.1 Å/min and (b) to a higher rate of 2.4 Å/min. (c-d) Topography and damping images (100 nm) corresponding to a detail of (a). Some islands with double-layer height and some of triple-layer height are labeled by numbers 2 and 3, respectively. Reprinted with permission from [Maier *et al.* (2007)]. Copyright (2007) by the American Physical Society.

investigated, and a quite different behavior was found. In such a case, the first NaCl layer stretched to the bulk lattice constant of the substrate, and the films was observed to grow flat. The dependence of the interface structure on the growth sequence can be attributed to the anharmonicity of the ionic bonds.

5.4 Oxide Films

The structure of ordered alumina films formed by oxidation of a NiAl(110) surface could be resolved by STM in 2005 [Kresse *et al.* (2005)]. The

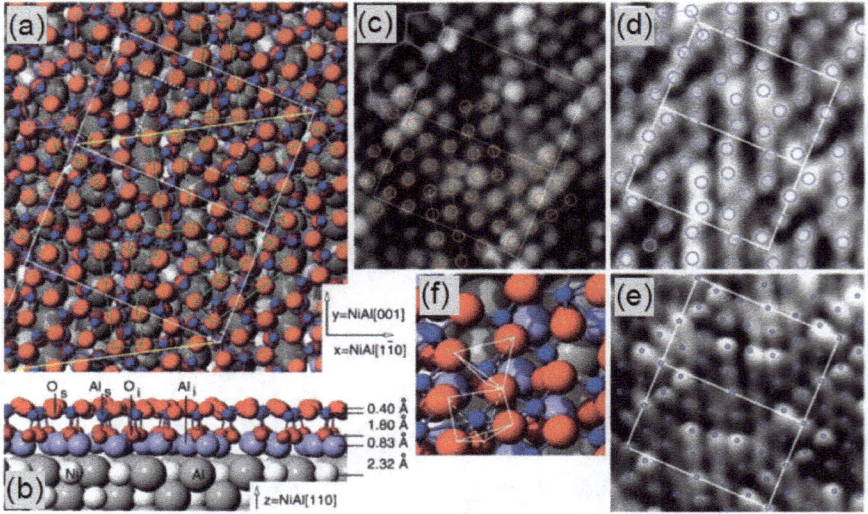

Fig. 5.14 (a) Top and (b) side view of a STM-based model for an ultrathin Al_2O_3 film formed on NiAl(110). (c-e) Experimental STM images of the film (c, d) at room temperature and (e) at low temperature. Two oxide unit cells are marked by white rectangles and the diagonal along which the oxide is commensurate is yellow. Green rectangles and squares highlight oxygen atoms in a square arrangement. Circles indicate the Al and O positions from (a) and (b) in the corresponding colors. (f) Close-up of the structure. From [Kresse *et al.* (2005)]. Reprinted with permission from AAAS.

square features on the surface in Fig. 5.14 (marked by green rectangles and squares) can be explained by the arrangement of the oxygen atoms. The stacking sequence and stoichiometry of the film is different from the commonly assumed Al_2O_3 stoichiometry. The oxide unit cell covers 16 NiAl surface unit cells and is commensurate to the substrate in the direction of the yellow diagonal of the unit cell. In the other direction, the film is incommensurate. The distance between neighbor Al atoms in the surface layer corresponds to the Al-Al distance in the surface layer of the reconstructed structure observed on bulk-truncated surfaces (Sec. 4.2). In another study, this structure was attributed to the dominant phase of the oxide film, and another reconstruction, characterized by zig-zag features arranged in parallel stripes, was also found [Gritschneder *et al.* (2007)]. On large scale, the secondary phase looked quite different from the dominant phase, whereas higher resolution images revealed great similarity of the atomic structures.

Ultrathin alumina films on $Ni_3Al(111)$ have been also studied. On these surfaces, STM images revealed two different hexagonal superstructures [Rosenhahn *et al.* (2000)]. One structure corresponded to the true crystalline mesh, which is commensurate with the Ni_3Al substrate. The other structure is a $(1/\sqrt{3} \times 1/\sqrt{3})R30°$ sublattice [Degen *et al.* (2005)]. Atomically resolved images of both structures could be obtained by NC-AFM [Hamm *et al.* (2006)]. Here the nodes of the two structures appeared pinned on points of the substrate lattice, where the surface atomic lattice is almost commensurable. The oxygen lattice had a perfect hexagonal symmetry close to these nodes but appeared disordered in the surrounding regions (Fig. 5.15).

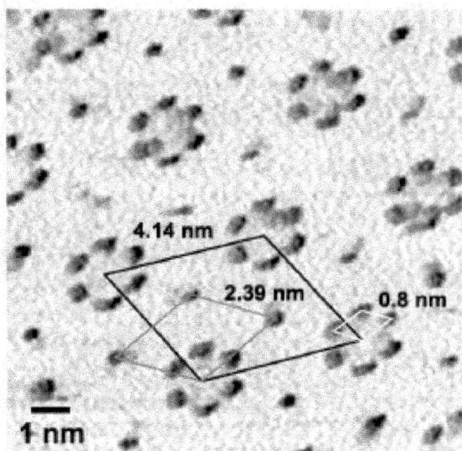

Fig. 5.15 Damping image of the alumina film formed after oxidation at 730 °C of a clean $Ni_3Al(111)$ surface. Three different hexagonal structures can be distinguished. Reprinted with permission from [Hamm *et al.* (2006)]. Copyright (2006) by the American Physical Society.

Scanning tunneling microscopy was also applied to ultrathin MgO films on Ag(001) at 225 °C [Schintke *et al.* (2001)]. Homogeneous square islands of 10-15 nm size appeared after deposition of 0.3 MgO monolayers [Fig. 5.16(a)]. Some islands close to a step edge were embedded in the upper terrace, which was attributed to the diffusion of Ag adatoms at high sample temperature. Depending on the bias voltage atomic resolution could be achieved on the MgO layer or, alternatively, on the Ag substrate (through the MgO layer). After deposition of about 2 MgO monolayers the substrate was completely covered with MgO, which formed terraces of typically 50 nm width and pyramidal islands [Fig. 5.16(d)]. Spectroscopy measurements on this system are discussed in Chapter 11.

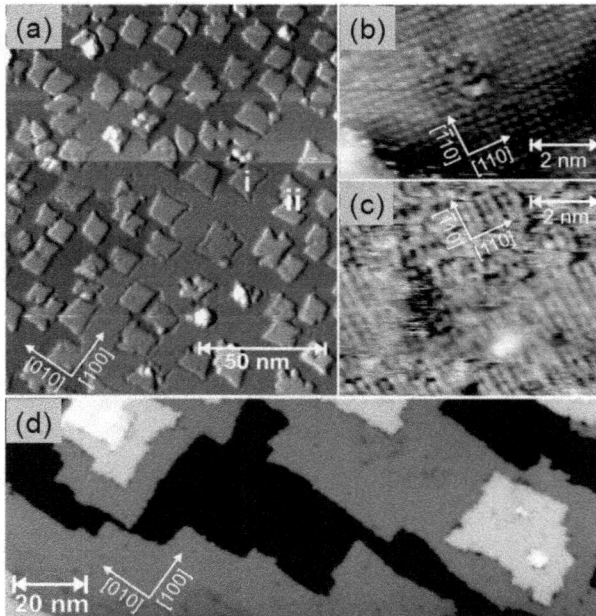

Fig. 5.16   STM images of (a) 0.3 ML MgO/Ag(001), (b) the Ag(001) substrate resolved through an MgO island, (c) the MgO layer atomically resolved (only one type of ions), and (d) 2.0 ML MgO/Ag(001). Reprinted with permission from [Schintke *et al.* (2001)]. Copyright (2001) by the American Physical Society.

The structure of a thin single crystalline SiO$_2$ film grown on Mo(112) was once again investigated by STM [Weissenrieder *et al.* (2005)]. Fig. 5.17 shows a film consisting of a 2D network of corner-sharing SiO$_4$ tetrahedra, with one oxygen of each tetrahedron binding to the protruding Mo atoms of the substrate. Complementary techniques revealed that, when the coverage exceeded two monolayers, the silica films showed properties comparable to bulk SiO$_2$ samples. The STM study was also supported by DFT calculations in excellent agreement with the experimental results.

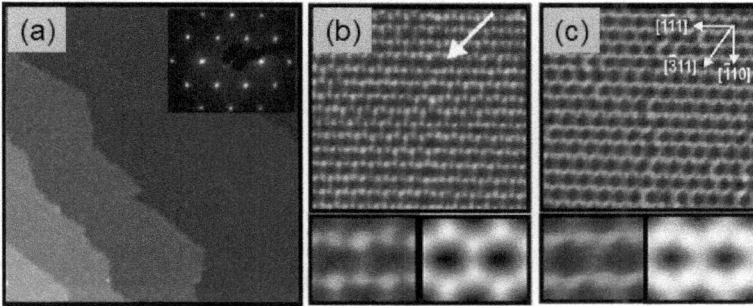

Fig. 5.17 STM images [(a) 100 nm, (b-c) 8 nm] of a $SiO_2/Mo(112)$ film. The arrow indicates an antiphase domain boundary running along the [$\bar{1}10$] direction. The insets in (b) and (c) show close-ups of the atomically resolved STM images (left) and simulated images based on DFT (right). Reprinted with permission from [Weissenrieder *et al.* (2005)]. Copyright (2005) by the American Physical Society.

5.5 Modeling AFM on Thin Insulating Films

The enhanced contrast observed by NC-AFM at step edges and kink sites of NaCl films grown on a copper surface (Sec. 5.1) could be explained by atomistic simulations [Bennewitz *et al.* (2000)]. As expected, the effect is due to the low coordination of the edge and kink ions, which quantitatively results in the following consequences. First, the electrostatic potential over the low coordination sites extends much farther than over the ideal terrace. Second, the low coordination of the ions increases their displacements due to the interaction with the probing tip. According to the simulations, the displacement is almost doubled for Na^+ ions at step edges and kink sites (0.24 and 0.25 Årespectively) compared to ions at the terraces (0.14 Å). In the case of kink sites, the position of several ions can be significantly modified, which explains the stronger contrast observed at such locations. In a similar way stronger contrast is expected on charged defects, such as impurities, adsorbed ions and vacancies. The same theoretical analysis also suggested that the NaCl film imaged by AFM did not exceeded a thickness of two monolayers.

Chapter 6

Interaction of Ions, Electrons and Photons with Halide Surfaces

The erosion of ionic crystal surfaces due to either *desorption induced by electronic transitions* (DIET), i.e. emission of surface particles due to the interaction of ionizing radiation with the electronic system of the solid, or due to *ballistic collisions* with momentum transfer between the projectile and the solid atoms, has been investigated for many decades [Townsend and Kelly (1968); Kelly and Lam (1973); Itoh (1976); Overeijnder *et al.* (1977); Szymonski (1993); Itoh and Stoneham (2000)]. The most striking feature of the experimental findings is that the *erosion yield* (number of desorbed/sputtered particles per incoming projectile) is unexpectedly large indicating very efficient transfer of the deposited energy into nuclear motion of the sample constituents. In particular, the yield of ion induced sputtering is considerably greater than one could expect from ballistic considerations only [Kelly and Lam (1973); Biersack and Santner (1982)].

6.1 Ion Bombardment of Alkali Halides

The slowing down of an ion projectile in a solid comes about by ballistic collisions (*nuclear stopping*) and by excitation and ionization of the crystal electronic system (*electronic stopping*). Therefore, the initial projectile energy E_0 can be split into two parts:

$$E_0 = \nu + \eta,$$

where ν corresponds to the fraction of initial energy lost in ballistic collisions and η is the fraction of initial energy lost for excitation and ionization. Sputtering processes occurring due to ballistic collisions with momentum transfer between the incoming projectile and the sample atoms were described by introducing a so-called *linear collision cascade model*

[Sigmund (1969)]. The model assumes that only binary collisions take place between the projectile and target atoms at rest and subsequently between the energetic recoiling atoms and other atoms within a cascade interaction volume. Under such assumptions the energy transfer in the cascade is a linear function of the initial projectile energy, E_0, and the cascade volume can be approximated by a Gaussian profile with well-defined longitudinal and transverse stragglings [Winterbon (1975)]. A simple formula was derived to describe the spectrum of translational energies of particles sputtered due to linear collision cascades, assuming a planar potential barrier for the recoiling atoms in the surface vicinity that received energy sufficient to overcome the binding to the solid [Thompson (1968)]. Following Thompson's formula the flux $\Phi_{coll}(E)$ of sputtered atoms leaving the surface with translational energy E is given by a simple equation:

$$\Phi_{coll}(E) \propto \frac{E}{(E + E_b)^3},\tag{6.1}$$

where E_b denotes a surface binding energy of the sputtered atom as a consequence of the binding force directed along the normal to the surface (*planar surface barrier*). In case of elementary metals the binding energy is well approximated by the sublimation energy. For compound ionic samples, however, the estimation of the binding energy is much more complicated. Furthermore, the assumption of planar surface barrier might not be valid too. A realistic estimate of the surface binding energy could be made experimentally by measurements of the translational energy distribution of sputtered particles of known mass. The energy corresponding to the position of the maximum flux determines a half value of the binding energy, E_b. Most often this has been accomplished by using time-of-flight method combined with mass analysis by means of a *quadrupole mass spectrometer* (QMS) [Szymonski *et al.* (1978); Szymonski and de Vries (1981)]. Another possibility invokes laser post-ionization of sputtered particles (resonant or non-resonant) combined with time-of-flight method [Henrich (1976); Hess *et al.* (2004)].

If the spatial density of the deposited energy in the collision cascade is very high, the collisions between both partners being in motion could take place, and sputtering cannot be described anymore by the linear model. As a result, the sputtered particles ejected in an early stage of the collision cascade could still obey the energy distribution given by (6.1), but at the later stage of the collision cascade collisions between moving partners take place, and the corresponding energy distribution is commonly described by a Maxwellian function:

$$\Phi_{\mathrm{non-lin}}(E) \propto E \exp\left(-\frac{E}{k_B T_{\mathrm{non-lin}}}\right), \tag{6.2}$$

where the parameter of the distribution $k_B T_{\mathrm{non-lin}}$ has a meaning of average energy lost by ballistic collisions per atom of the interaction volume [Szymonski (1982); Sigmund and Szymonski (1984)]. In some descriptions of non-linear sputtering a so-called *thermal spike model* was used [Thompson and Nelson (1962)], and the $T_{\mathrm{non-lin}}$ parameter was interpreted as an equilibrium temperature of the cascade volume, although in case of alkali halides a solid theoretical treatment of such phenomena is not available.

The first systematic experimental energy distributions for alkali and halogen atoms sputtered from alkali halides with an ion beam were measured by the end of the 1970's [Szymonski *et al.* (1978); Szymonski and de Vries (1981); Overeijnder *et al.* (1978b)]. The energy distribution of iodine atoms sputtered from a LiI sample using Kr ions is shown in Fig. 6.1. Many important features of the sputtering process could be established, although all the work was done for compressed powders having presumably polycrystalline structure (not verified experimentally). There are some indications (see below) that in the case of single crystals, transport processes of local excitation products are much more efficient than for amorphous or polycrystalline samples. Such phenomena could affect electronic desorption/sputtering processes and new channels for particle emission could be opened [Itoh and Stoneham (2000)].

In fact, it was found that 15 keV Ar ion bombardment of a NaCl single crystal resulted in emission of only thermal particles in contrast to energy distribution measurements for compressed powders, where both ballistic and thermal components had been found [Yu *et al.* (1981)]. This finding strongly indicated that in the case of single crystals electronic sputtering is by far more efficient than for polycrystalline/amorphous samples. It is striking that a dominating part of the sputtered flux consisted of particles precisely described by a *thermal* (Maxwellian) distribution of translational velocities. Such spectrum converted into energy distribution reads as:

$$\Phi_{\mathrm{th}}(E) \propto E \exp\left(-\frac{E}{k_B T}\right), \tag{6.3}$$

where T is the macroscopic temperature of the irradiated surface. It was shown that this thermal contribution to sputtering scales with the projectile electronic stopping and resembles a new mechanism specific to

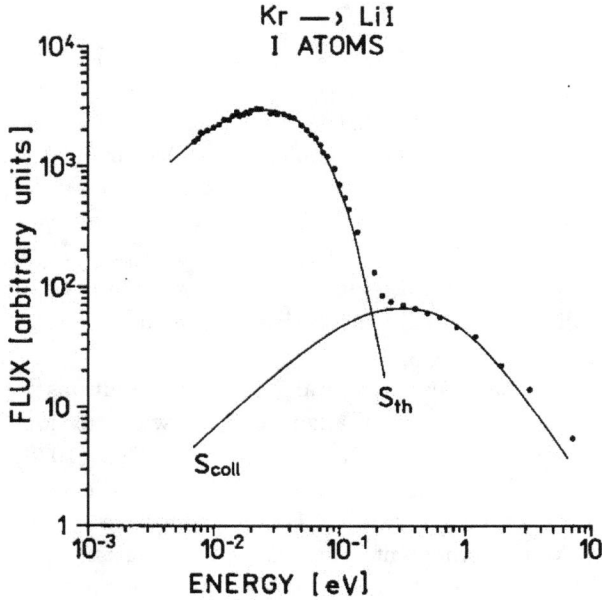

Fig. 6.1 Energy distribution of I atoms sputtered from LiI with 6 keV Kr ions. The ballistic and electronic contributions to sputtering obtained by fitting Eqs. (6.1) and (6.3) to the experimental points are denoted by S_{coll} and S_{th} respectively. From [Szymonski and de Vries (1981)]. Reprinted by permission of the publisher (Taylor & Francis Group).

ionic crystals, in which the energy deposited initially in the electronic system of the crystal, $\eta(E_0)$, is eventually converted into motion of the decomposed crystal constituents, such as alkali and halogen neutral atoms emitted from the surface with thermal spectrum of energies [Krok *et al.* (2004a)] (see Fig. 6.2).

6.2   Electron and Photon Stimulated Desorption

One can study the electronic process of sputtering (often called 'desorption', although it concerns native constituents of the crystal, not surface adsorbates) more precisely by using electrons or energetic photons as projectiles. In such cases the momentum transfer in ballistic-type collisions is eliminated, and only surface dynamics due to electronic transitions takes place. For alkali halides it was found that the atomic constituents of the irradiated crystal are emitted as desorbed species. Although the

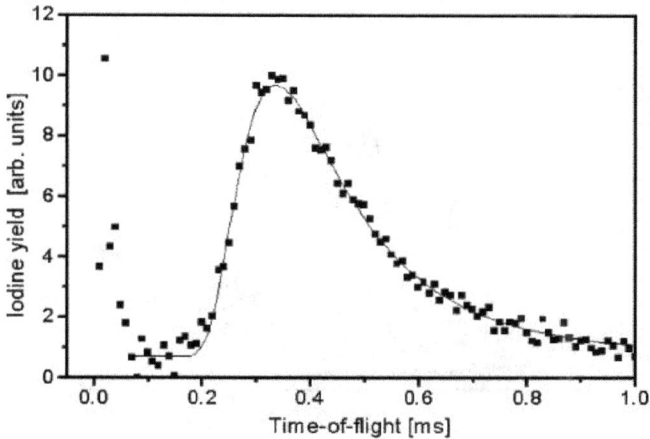

Fig. 6.2   A time-of-flight distribution of I atoms emitted at an angle of 45° from RbI(001) due to bombardment with 5 keV He$^+$ at about 115 °C. The solid line represents the best fit of Maxwellian distribution with temperature 125 °C. Reprinted from [Krok *et al.* (2004a)], Copyright 2004, with permission from Elsevier.

efficiency of the desorption process depends on the experimental conditions, in general, the flux of emitted particles consists of 80-90% of neutral atoms, 10-20% of molecules and a very small fraction ($\sim 10^{-3}$-10^{-5}) of ions [Szymonski (1993)].

6.2.1   *Electron stimulated desorption*

Translational energy distributions of alkali and halogen atoms desorbed from alkali halide surfaces with low energy electrons (200-600 eV) were measured for compressed powder samples [Overeijnder *et al.* (1978a,c)], for single crystal KBr [Postawa and Szymonski (1989)] and for several other alkali halides [Kolodziej *et al.* (1994)]. A set of time-of-flight spectra obtained for various sample temperatures is shown in Fig. 6.3. In general, the distribution consists of exclusively thermal (Maxwellian) component for alkali atoms, and two components: thermal and *non-thermal* ones (with a maximum around 0.2-0.3 eV) for halogen atoms. The thermal contribution to desorption is strongly temperature dependent and, therefore, the relative rate of non-thermal to thermal components in halogen atom spectra decreases rapidly with temperature.

700 eV e⁻ on (100) KBr

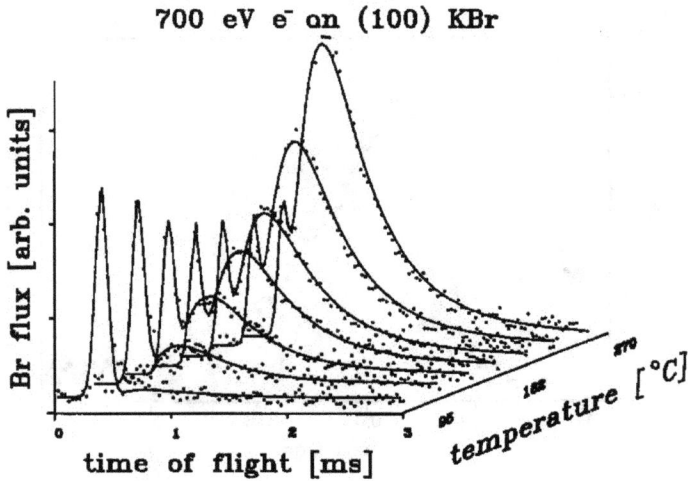

Fig. 6.3   Time-of-flight distributions of Br atoms leaving a KBr(001) target at various surface temperatures.

The angular distribution of desorbed particles is strongly forward peaked [for the (001) surface] for non-thermal halogen atoms and isotropic (cosine-like) for thermal atoms of both crystal components [Postawa and Szymonski (1989); Szymonski *et al.* (1991)]. These findings have several implications for understanding electronic desorption mechanisms of alkali halides.

6.2.2   *Photon stimulated desorption*

Most of the experimental observations, so far, yielded qualitatively similar results for both *electron-* (ESD) and *photon-* (PSD) *stimulated desorption*. There are several reports, however, which emphasize important differences between the two excitation paths [Hess *et al.* (2005); Wurz *et al.* (1991); Liu *et al.* (1993)]. In particular, recent energy distribution measurements for halogen atoms desorbed due to UV-laser irradiation with photon energies close to the threshold (the fundamental *surface exciton*) found predominantly the non-thermal distribution, as seen in Fig. 6.4 [Hess *et al.* (2004)]. Both thermal and non-thermal velocity components were found for photon excitations with energies higher than the band gap energy. This

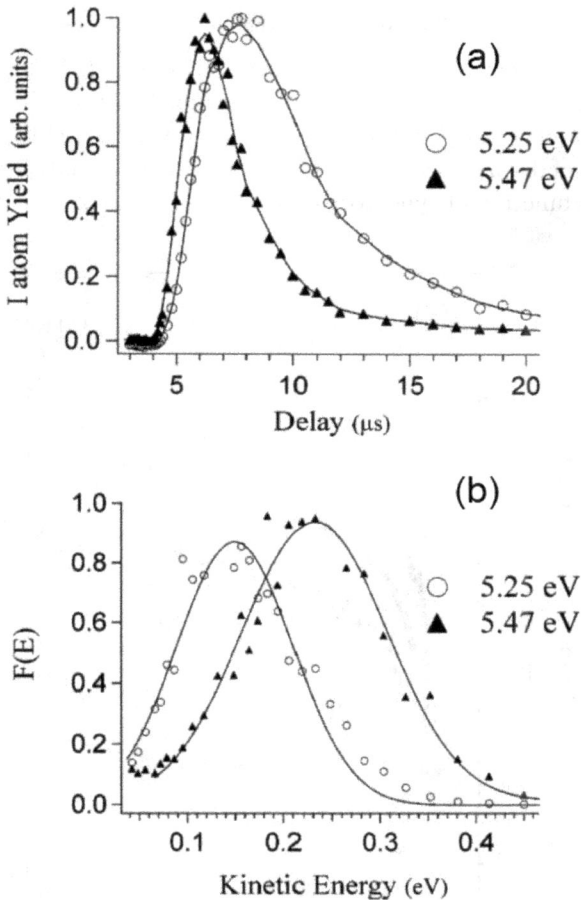

Fig. 6.4 (a) Velocity profiles of non-thermal $I(^2P_{3/2})$ atoms desorbed from KI single crystal by 5.47 eV and 5.25 eV laser excitation. The velocity profile is obtained by recording the relative I yield as a function of a delay time between pump and probed lasers. The corresponding kinetic energy distributions are shown in (b). Reprinted from [Hess *et al.* (2004)], Copyright 2004, with permission from Elsevier.

finding may suggest that at threshold excitation energies only a mechanism leading to emission of non-thermal particles is activated, whereas for excitation energies allowing for electron-hole pair creation a new efficient mechanism leading to thermal desorption is possible.

6.2.2.1 *Desorption by excitation at threshold energies*

The yield of photon-stimulated desorption induced by excitation at threshold energies (5-10 eV) was first studied for a KI(001) sample [Brinciotti *et al.* (1991, 1994)]. Relative desorption yields and angular distributions of desorbing atoms were also measured for (001) surfaces of KI and RbI as a function of synchrotron radiation energy [Zema *et al.* (1997)].

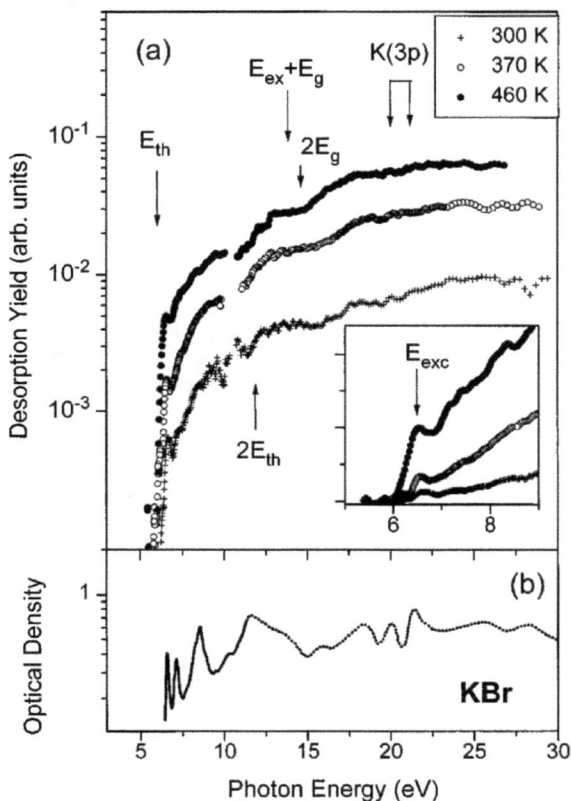

Fig. 6.5   (a) I atom desorption signal measured as a function of synchrotron radiation energy irradiating a KI(001) surface at different sample temperatures. The arrows mark the threshold energies for electron-electron scattering and the position of the K^+ $3p$ core excitons. The inset shows the desorption yield in the threshold region. The arrow in the inset shows the position of the first exciton. (b) For comparison an absorption spectrum of KI measured at room temperature is taken from [Ejiri (1987)] (solid line) and [Eby *et al.* (1959)] (dashed line). Reprinted with permission from [Zema *et al.* (1997)]. Copyright (1997) by the American Physical Society.

An example of such dependence for I atoms desorbing from RbI(001) crystal is presented in Fig. 6.5. For comparison, the surface reflectivity dependence on the synchrotron radiation energy measured in parallel with the PSD signal is displayed in the same plot. It is clear that the onset of the desorption occurs at an energy corresponding to the lowest excitonic state within the band gap energy of the crystal.

More recently, PSD threshold experiments with UV-laser excitation have been also reported [Hess *et al.* (2001, 2002); Henyk *et al.* (2003); Hess *et al.* (2004); Beck *et al.* (2004); Hess *et al.* (2005)]. Thanks to the high selectivity of the laser excitation it was possible to identify the desorption yield onset corresponding to the Urbach tail in the excitation function, i.e. in the range of the surface exciton creation. Based on the assumption that a non-radiative decay of the surface exciton results in non-thermal halogen emission [Hess *et al.* (2004); Beck *et al.* (2004)], experimental values of the surface exciton energy shifts with respect to the bulk exciton values could be derived for various alkali halides (Fig. 6.6).

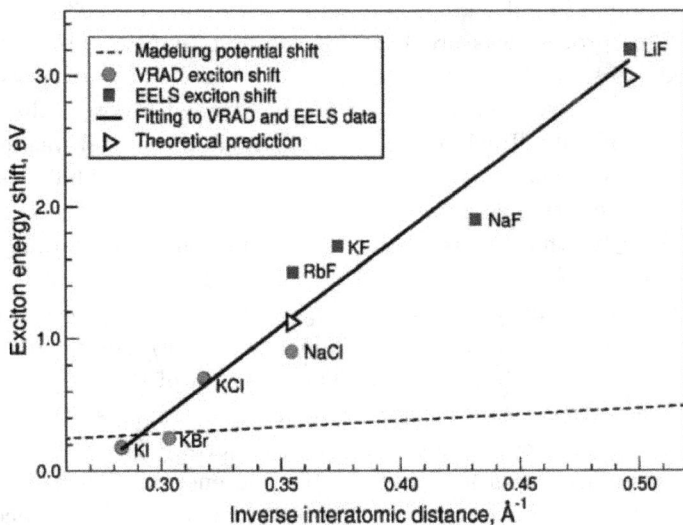

Fig. 6.6 Surface exciton energy shift, ΔE, with respect to bulk exciton energy (in eV) versus the inverse interatomic crystal lattice spacing (in Å^{-1}) for several alkali halides. The open triangles show ΔE calculated theoretically for NaCl and LiF. The solid line represents the best linear fit to the experimental points and the dotted line displays the Madelung potential contribution to the surface exciton shift. Reprinted with permission from from [Hess *et al.* (2005)]. Copyright 2005 American Chemical Society.

For the particular case of RbI and KI the energy difference between the surface and the bulk excitons is rather small, thus threshold excitation with less selective sources (such as electrons or synchrotron radiation) leads to the creation of both surface and bulk excitons. This is consistent with most recent mass selected time-of-flight measurements indicating that I atoms desorbed from RbI with deuterium lamp UV-photons contain both velocity components, i.e. thermal (Maxwellian) and non-thermal ones [Szymonski *et al.* (2006)].

6.2.2.2 *Desorption due to band-band excitation*

For excitation energies above the band gap value pairs of conduction band electrons and valence band holes are created. Since the width of the valence band is quite large (of the order of 3 eV for KBr) the holes can be quite energetic (*hot holes*) with their excess energy allowing for high mobility within the crystal and large diffusion lengths [Elango (1994)]. The surface recombination of the pair occurs with prompt emission of non-thermal halogen and creation of a *surface F-center* [Szymonski (1993); Hess *et al.* (2005)]. This process was proposed for the first time to explain photon-stimulated ejection of halogen atoms from alkali halide nanocrystals [Li *et al.* (1992)]. Later such a process was confirmed theoretically for the (100) surface of NaCl [Puchin *et al.* (1994)]. A similar model was proposed to explain non-thermal emission of halogen atoms from alkali halide surfaces at threshold excitation energies (Fig. 6.7).

Alternatively, the electron-hole pair could self-trap, leading to the production of *Frenkel pairs* (so called F-center and *H-center* pair of defects, where H denotes a halogen atom in the interstitial position and F stands, as usual, for an electron trapped by the halogen vacancy) [Song and Williams (1993); Itoh and Stoneham (2000)]. The diffusion of these defects in the vicinity of the surface can lead to thermal desorption of crystal constituents (see Fig. 6.8). A H-center migrating to the surface produces a halogen adatom [Szymonski (1980)]. Since the binding energy of such an atom is relatively low (0.14 eV [Puchin *et al.* (1993)]), it leaves the surface easily even at room temperature. The behavior of an F-center in the vicinity of the crystal surface is less known. In fact, during electron irradiation of alkali halides (NaCl(001) crystal [Kubo *et al.* (1994)]) F-centers in the ground state are accumulated under the crystal surface. It was demonstrated that ground state F-centers in the bulk of the alkali halides are immobile, but those excited (by light or by irradiating electron beam) become highly

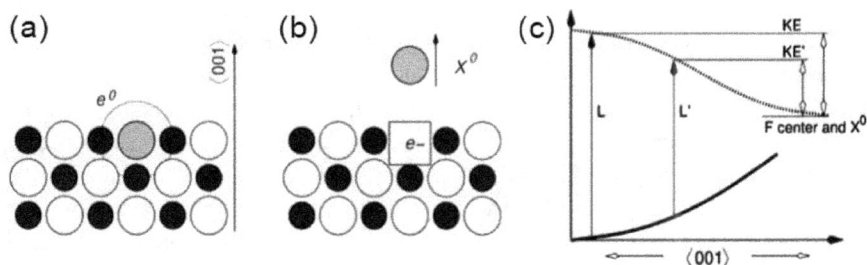

Fig. 6.7   Model of surface exciton decomposition and halogen-atom desorption from an alkali halide surface. Schematic of (a) the localized surface exciton and (b) displacement of the neutral halogen atom (X^0) perpendicular to the surface due to surface exciton decomposition. The lower curve (c) is the adiabatic potential of the crystal ground state with respect to the same $\langle 001 \rangle$ symmetry axis. The arrows indicate vertical excitation in the near and far tail region using lasers of photon energies L and L'. This model predicts X^0 desorption as shown schematically in (b) and significant reduction in X^0 kinetic energies (KE$'$ < KE) as the laser photon energy is decreased ($L' < L$). Reprinted from [Hess et al. (2004)], Copyright 2004, with permission from Elsevier.

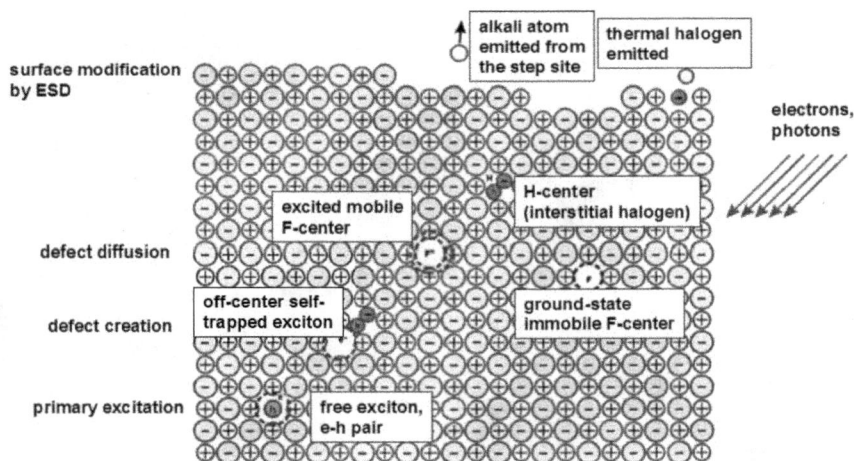

Fig. 6.8   Schematic diagram of DIET processes in alkali halides leading to thermal emission.

mobile [Salminen et al. (1996)]. Therefore, illumination of the electron pre-irradiated crystal with light at a wavelength corresponding to the F-center absorption band [Kubo et al. (1994)] drives the accumulated F-centers to the surface and causes desorption of alkali atoms.

Theoretical adiabatic potential energy surface (APES) calculations showed that an F-center in the ground state cannot initiate desorption because of an energy deficit of about 2 eV [Puchin *et al.* (1994)]. Furthermore, a 2*p*-excited F-center could have sufficient energy to desorb an atom but there is an energy barrier which inhibits the desorption. Desorption is more probable, however, from low-coordinated surface lattice sites like steps, kinks and corners. The accumulation of ground-state F-centers together with the preferential desorption from low-coordinated sites on the surface plays a crucial role in explanation of the desorption yield oscillations as a function of the absorbed dose of electrons [Such *et al.* (2000b); Puchin *et al.* (1994); Such *et al.* (2000a)] (see also Chapter 7).

Chapter 7

Surface Patterning with Electrons and Photons

As shown in the previous chapter, excitation by electrons or photons results in the desorption of atoms from alkali halide surfaces. At the first stages of the typically layer-by-layer surface removal, monatomic-depth rectangular pits are formed, which are several nanometers in size. These pits can be used to trap small numbers of organic molecules, which is an important step towards assembling a molecular electronic device on an insulating substrate or, alternatively, they may help to control the nucleation of metal nanostructures at surfaces, as discussed in Chapters 9 and 10. On insulating oxides, particle emission under electron irradiation was observed as a transient effect due to bond breaking following by oxygen emission [Kelly (1979)]. However, this process is not continuous and no noticeable surface erosion could be recognized, since the metal or metal-enriched surface layer prevented any further desorption.

7.1 Surface Topography Modification by Electronic Excitations

7.1.1 *Layer-by-layer desorption*

In an early study, the NaCl(001) surface desorption due to UV-photons generated by a deuterium lamp was monitored using He atom scattering [Hoeche *et al.* (1994)]. Characteristic oscillations of the scattered He intensity were found, indicating a *layer-by-layer* mode of desorption (Fig. 7.1). Furthermore, it was found that such a mode of erosion occurred in a wide range of crystal temperatures (0-220 °C). However, the mass spectrometric measurements of the desorption signal performed in the same work did not indicate any dependence of the desorption yield on the surface topography. In a more recent work NC-AFM was used for imaging surfaces desorbed under similar experimental conditions (UV deuterium lamp excitation, same sample temperature range) [Szymonski *et al.* (2002)]. For monitoring the desorption signals a QMS, as well as a sensitive surface ionization detector for the alkali atoms, were used. The measurements confirmed the previous findings that the desorption occurs in a layer-by-layer mode.

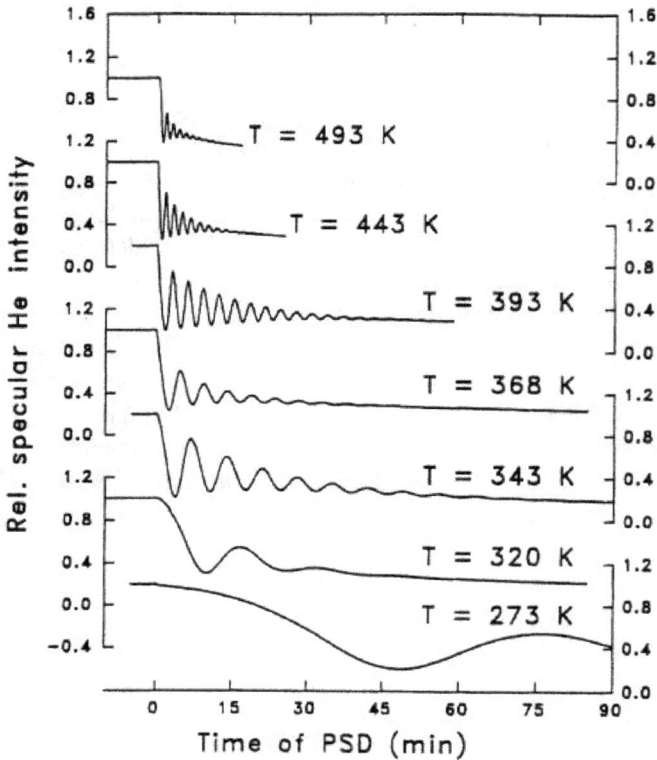

Fig. 7.1   Helium-atom scattering signal for NaCl(001) irradiated with UV photons. Reprinted with permission from [Hoeche *et al.* (1994)]. Copyright (1994) by the American Physical Society.

It is important to note that the yield oscillations are also found for electron irradiated alkali halides [Such *et al.* (2000b); Szymonski *et al.* (2001)] and for both desorption components, i.e. for alkali and halogen atoms, as seen in Fig. 7.2. In the case of photon irradiated RbI(001) [Szymonski *et al.* (2006)], since the average UV light absorption coefficient at 400 K can be evaluated as equal to 3×10^5 cm^{-1} [Peimann and Skibowski (1971)], it could be assumed that most of the primary photon energy is deposited down to a depth of 60 nm, which is similar to the penetration depth of 1 keV electrons [Bronshteyn and Protsenko (1970)]. This strongly indicates that the effect of surface topography is not dependent on the primary excitation details but it is rather affecting the yield in later stages of the process when diffusion and/or surface recombination of the secondary excitation products take place.

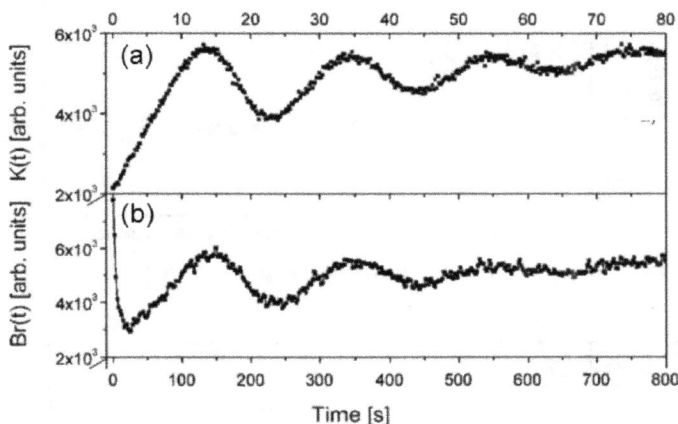

Fig. 7.2 Yield oscillations corresponding to layer-by-layer erosion of the KBr(001) surface. Reprinted with permission from [Such *et al.* (2000b)]. Copyright (2000) by the American Physical Society.

The analysis of the oscillatory dependence of the desorbing atom fluxes on the irradiation time may be used for estimation of the desorption rate. Since the removal of a single monolayer corresponds to a full period of the oscillation, one could calculate the desorption yield even in absolute units, providing that the accurate number of impinging photons or electrons per irradiates area is known. Such analysis was performed for NaCl(001) ESD yield dependences measured at various temperatures of the sample [Szymonski *et al.* (2001)]. The results are shown in Fig. 7.3, where the dependence of the logarithm of the inverse period (which is proportional to the efficiency of desorption) on the inverse of the sample temperature is shown. The calculated total yield dependence on the sample temperature is shown too. A section of the dependence can be fitted accurately by a straight line. The slope of the line corresponds to an activation energy of 0.25±0.2 eV, which agrees well with previous results [Hoeche *et al.* (1994)].

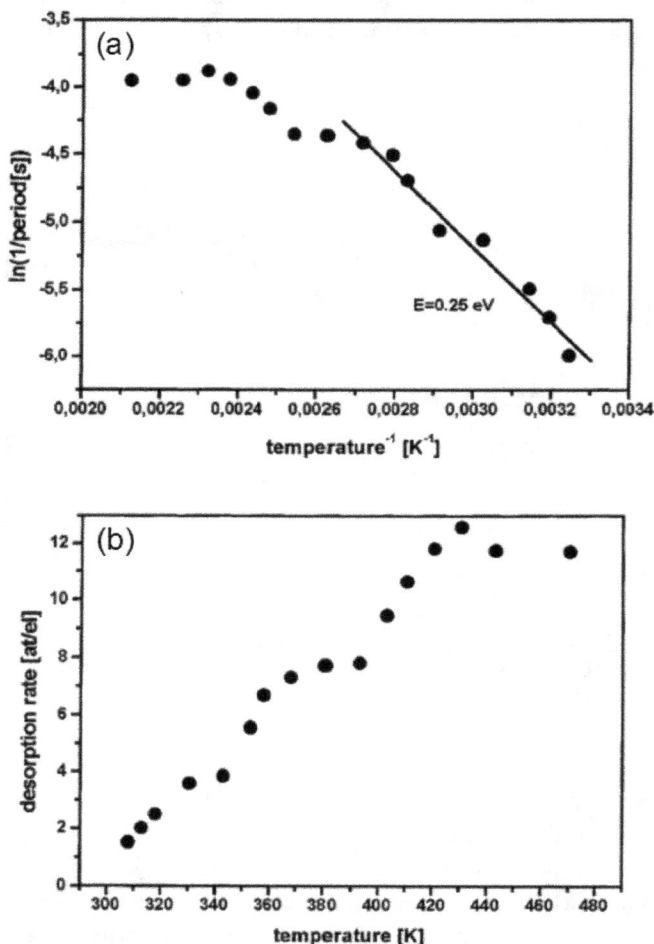

Fig. 7.3 1 keV electron irradiation of NaCl(001) at various temperatures of sample. (a) Dependence of the logarithm of the inverse period on the inverse of sample temperature. (b) Absolute yield dependence on temperature. The slope of the line corresponds to an activation energy of 0.25±0.2 eV. Reprinted from [Szymonski *et al.* (2001)], Copyright 2001, with permission from Elsevier.

7.1.2   Coexcitation with visible light

In order to better understand the role of electronic excitation of the
F-centers in photon stimulated desorption, a series of PSD experiments with
simultaneous excitation by UV and visible light was performed [Szymonski
et al. (2006)]. In Fig. 7.4 the yield of Rb atoms emitted from the RbI surface
at initial stages of UV-photon irradiation is presented. After removal of 1.5
monolayers (i.e. in the maximum of the second cycle of oscillations of the
yield) the visible light was turned on. As the result, an instantaneous
desorption signal increase was recorded, as well as a decrease of the period
of the oscillatory dependence. Both these effects occurred as a result of
considerable increase of the desorption yield.

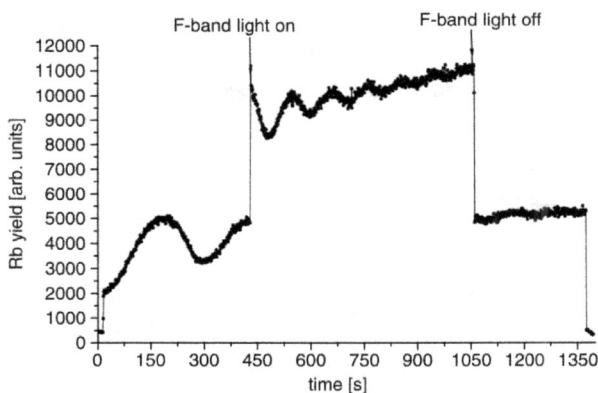

Fig. 7.4   Dependence of Rb desorption signal on time of UV-photon irradiation for
RbI(001) surface and simultaneous co-excitation by F-band light. The moments of
turning on and off F-band light are shown by arrows. From [Szymonski *et al.* (2006)].
Reproduced with permission from IOP Publishing.

Furthermore, it was found that additional illumination of the sample
with light causes significant increase of the desorption yield for both halogen
and alkali atom components [Szymonski *et al.* (2006)]. Although the
magnitude of the signal increase is similar for both components, there are
important differences in the character of 'light-on' and 'light-off' changes of
the signals. Whereas the Rb yield rises sharply at first and then decreases to
almost half of the initial increase, the I signal (not shown) rises more slowly
approaching the saturation value only after several seconds of irradiation.
Switching off the visible light causes a rapid fall of the Rb signal followed by

some increase into a saturation level and slower decrease of the I signal. The same behavior was observed for light co-excitation in electron stimulated desorption. Using a set of interference filters it was possible to measure a relative yield enhancement as a function of the wavelength of the visible light. The experimental studies performed for KBr crystals indicated that a relative ESD yield enhancement as a function of the wavelength of the visible light follows the F-center absorption curve for this crystal (Fig. 7.5). This clearly shows that the $2p$ excited F-centers are significantly more mobile than the ground state ones, and this enhanced mobility facilitates the emission of both desorption components. In order to understand the nature of F-center diffusion one has to remember that the $2p$-excited state of the F-center is located within the unoccupied states of the conduction band of the crystal. Therefore, it is likely that the lifetime of such excited F-center is rather short. On the other hand, the $2p$ excited center might be considered as equivalent to an electron in the conduction band and a halogen vacancy which undergo uncorrelated migration in the crystal.

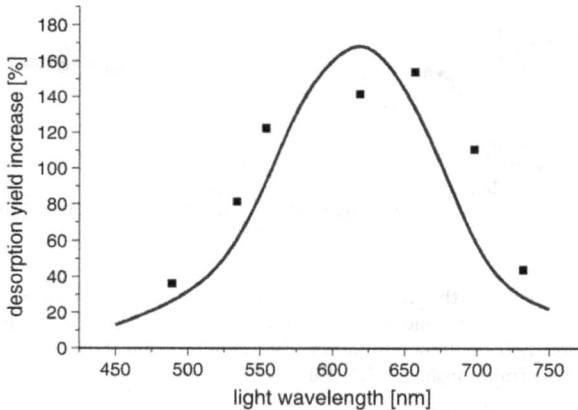

Fig. 7.5 Spectral dependence of the relative K desorption yield increase induced by visible light illumination of the electron bombarded KBr crystal. The solid line is drawn to guide the eye and represents the optical absorption profile of KBr F-centers. From [Szymonski *et al.* (2006)]. Reproduced with permission from IOP Publishing.

Although it is likely to expect that the excited F-center arriving at the surface can cause neutralization of surface alkali ion which then could evaporate thermally, the enhancement of the halogen component requires more consideration. Apparently, the enhanced emission of alkalis in expense of stable ground state F-center concentration is linked to the increase of the halogen signal. In a steady state, the H-center diffusion takes place in a lattice containing a number of ground state F-centers which can act as annihilation traps. Furthermore, the concentration of stable F-centers is controlled by the surface topography. As a result of F-center and H-center annihilation the thermal halogen signal should vary with the irradiation dose, 'sensing' the inverse F-center concentration at surface proximity. Such behavior was demonstrated for electron stimulated desorption [Such *et al.* (2000b)].

7.2 Nanoscale Pits on Alkali Halide Surfaces

Initially, as the result of electron stimulated desorption of alkali halides, the crystal surface is covered by *rectangular pits* in the first ionic plane of the sample with edges oriented along the main crystallographic directions ($\langle 100 \rangle$ and $\langle 010 \rangle$). The pits are randomly distributed, with an average size increasing with increasing electron fluence. An example of such a surface imaged with NC-AFM is shown in Fig. 7.6(a). The image was obtained for KBr(001) irradiated with 1 keV electrons at the surface temperature of 135 °C. The NC-AFM technique allowed for atomic resolution imaging of such structures, as seen in Fig. 7.6(b). It is interesting to note that the edges of the pits are often not following a straight line but have one or more single-atom kinks, which represent the sites with the lowest coordination number. It is likely that such sites (steps and kinks) could act as the precursor sites for defect trapping and emission of desorbing particles.

A further increase of the electron fluence resulted in enlargement and overlap of the pits. The surface topography varied from initially flat to rough, with the maximum number of low-coordinated sites for 1/2 ML eroded, and back to flat for a fluence corresponding to the removal of a complete monolayer. One should note that the areas inside the pits are undamaged and atomically flat for the first period of desorption [Fig. 7.6(b)] and there is no noticeable time dependence of this topography for several hours or even days, provided that the samples are kept in the UHV environment. A possibility of healing of the radiation damage inside the pits due to thermal diffusion in the time period between irradiation and imaging could not be excluded based on AFM image analysis alone.

The topography variation of the irradiated surface is closely correlated with the oscillations of the desorption yield detected 'in situ' with a

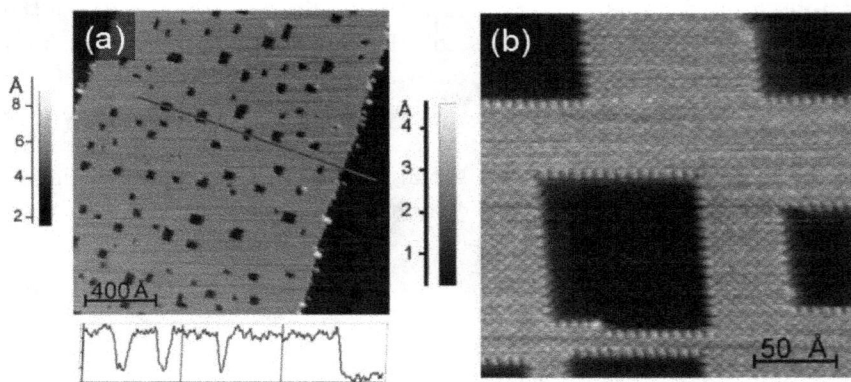

Fig. 7.6   (a) NC-AFM image (180 nm) of a KBr(001) surface irradiated with low fluence of electrons (approx. 1 μC/cm^2). A cross section is shown along the marked line in order to demonstrate that the level difference between the undamaged surface and the bottom of the pits corresponds to the distance between two neighboring crystal planes in KBr. (b) Atomically resolved NC-AFM image (23 nm) of irradiation-induced pits in a KBr surface. The depth of the pits is equal to the distance between two neighboring crystal planes. Reprinted from [Goryl *et al.* (2005)], Copyright 2005, with permission from Elsevier.

QMS (Fig. 7.7). The yield of both desorption components (alkali and halogen) oscillates with the electron fluence until the coherence is lost and the surface becomes covered by a multilevel network of pits and islands. Therefore, the AFM images of irradiated crystals represent a true surface morphology development predominantly caused by desorption process alone. Furthermore, the average distance between neighboring pits may be associated with a diffusion length of the precursor-state responsible for desorption. Also, if the size of the pit becomes comparable to this diffusion length, new pits might be initiated inside the old ones [Kolodziej *et al.* (2001)]. A similar behavior was found for photon stimulated desorption (Fig. 7.8).

It was also shown that only a fraction of the atoms emitted with thermal (Maxwellian) distribution of velocities oscillates, i.e. it is sensitive to the surface topography variations [Szymonski *et al.* (2001)]. This behavior is common for both the halogen and the alkali thermal components. In contrast, the emission efficiency of non-thermal halogen atoms is not sensitive to variations of the surface topography. This is consistent with the model of non-thermal halogen emission described in Chapter 6.

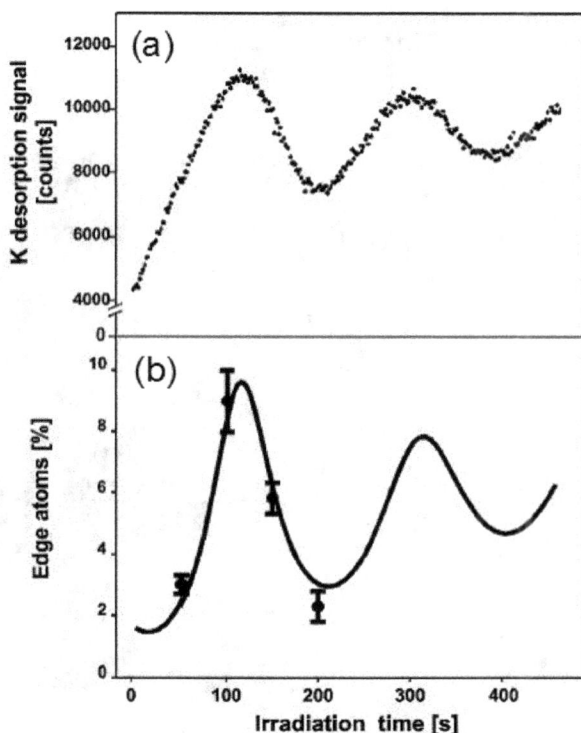

Fig. 7.7 (a) Potassium desorption yield from electron irradiated KBr crystal, as measured with a quadrupole mass spectrometer. (b) Surface recombination coefficient calculated from the equation described in the text and scaled to fit the experimental points obtained from NC-AFM images expressed as the percentage of surface edge atom density. Reprinted from [Goryl *et al.* (2005)], Copyright 2005, with permission from Elsevier.

7.2.1 *Diffusion equation for F-centers*

A simple diffusion equation with varying boundary conditions was used to describe the oscillatory yield dependence of both (alkali and halogen) thermal components of desorption and its correlation with the oscillating morphology of the irradiated surface area [Goryl *et al.* (2005)]. Since Frenkel defects are created in pairs and they are steadily supplied to the crystal by irradiation, the characteristic times for diffusion of defects are a few orders of magnitude smaller than the time required for significant change of the surface topography (milliseconds versus seconds). Therefore,

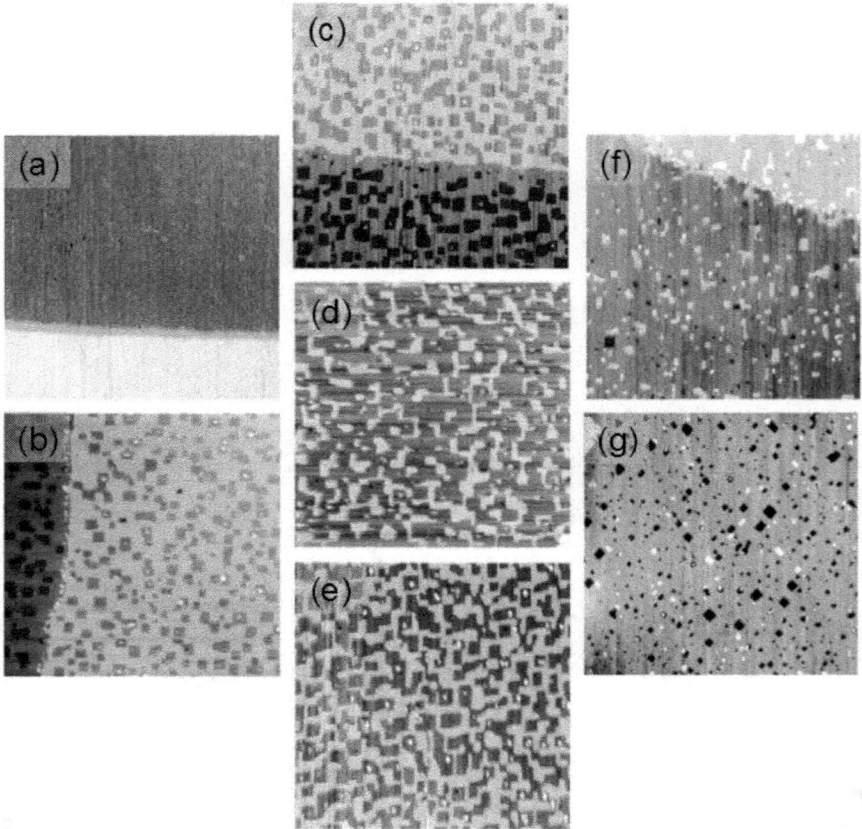

Fig. 7.8 A set of NC-AFM images (0.45 μm) illustrating the evolution of a RbI(001) surface irradiated with UV-light ($\lambda \leq 200$ nm) at a sample temperature of 100 °C. The sequence starts from a virgin surface (a) and continues with gradually increasing the photon dose corresponding to the yield oscillations shown at the bottom: (c) corresponds to the yield maximum of the first cycle, (e) corresponds to the yield minimum after the first cycle, (b) and (d) correspond to intermediate irradiation times. (f) represents the second cycle of the RbI surface evolution: the first topmost plane is almost totally desorbed and pits are grown in the next atomic plane. From [Szymonski *et al.* (2006)]. Reproduced with permission from IOP Publishing.

a steady-state diffusion equation can be used to describe the migration of the excited F-centers:

$$D\frac{\mathrm{d}^2 n(x)}{\mathrm{d}x^2} - \frac{n(x)}{\tau} + C(x) = 0, \tag{7.1}$$

where $n(x)$ is the F-center density in a layer parallel to the (001) surface at depth x, D is the diffusion coefficient, τ is the lifetime of the F-centers, and $C(x)$ is the density of energy deposited in the crystal at the depth x. It is plausible to assume that the rate of creation of excited F-centers is proportional to the density of energy deposition, which, in turn, can be approximated by two line segments [Jammal *et al.* (1973)]: the first one starting from a finite value of the deposited energy at the surface, $C(0) = C_{surf}$, connecting it with a maximum value of $C(x_{max}) = C_{max}$, and the second one continuing from C_{max} to $C(x_{range}) = 0$, where x_{range} is the total range of impinging electrons. With this approximation the desorption equation can be solved analytically:

$$n(x) = A_1 e^{x/\lambda} + A_2 e^{-x/\lambda} + \tau C(x),$$

where A_1 and A_2 are constant for each of the two segments of $C(x)$ and $\lambda = \sqrt{D\tau}$ is the diffusion length for excited F-centers.

It is plausible to assume that $n(x)$ is a smooth function. In order to fulfill that condition the first derivatives at the ends of linear ranges of $C(x)$ must be equal. Therefore, the constants A_1 and A_2 can be found. As for boundary conditions, the behavior of the solution at $x = 0$ and $x = d$, where d is much bigger than the range of the impinging electrons, must be considered. The F-center reaching the surface can either induce desorption of an alkali atom, if localized at a low-coordinated site, or it can be reflected back to the bulk when an atomically flat surface area was hit. Therefore, we postulate that the flux of F-centers recombining with the surface (and the flux of desorbed alkali atoms) is proportional to the concentration of F-centers at the surface: $J(\Delta_0) = \Delta_0 n(0)$, where Δ_0 is the surface recombination velocity which accounts for the probability of F-center trapping and desorbing at the surface per time unit. We expect that the desorption flux is large for large Δ_0 ($\Delta_0 \rightarrow \infty$ for perfect trap surface) and small for $\Delta_0 \rightarrow 0$ (the surface is a perfect mirror). Since the creation of defects at the boundaries of the migration volume is negligible, we arrive at the following boundary conditions for Eq. (7.1):

$$D\frac{dn(0)}{dx} = \Delta_0 n(0) \quad \text{and} \quad D\frac{dn(d)}{dx} = \Delta_0 n(d).$$

Assuming that the boundary lies deeply in the crystal, it is plausible to put $\Delta_d = 0$. The range of the impinging electrons and the diffusion length of the defects are much smaller than the width of the crystal (typically 2-3 mm in the experiments). Hence, it is unlikely that any defect can reach the rear face of the crystal and cause desorption, thus, the face can be assumed as being perfectly reflecting. For practical solution of the diffusion equation the value of the radiative life time of excited F-center in KBr (1.1 μs [Swank and Brown (1963)]) and the diffusion coefficient from [Salminen *et al.* (1996)], 2 nm^2/ns, can be used. Thus, the dependence of the desorption flux on the surface recombination coefficient can be calculated from

$$J(\Delta_0) = \Delta_0 n(0) = \frac{a\Delta_0}{b + \Delta_0}, \tag{7.2}$$

where the parameters obtained using the data mentioned above are $a = 127$ and $b = 0.88$.

The surface recombination coefficient is the factor that governs the efficiency of the desorption process and according to the model presented above should be proportional to the density of low-coordinated sites on the surface. To verify that claim the data obtained in previous experiments were used [Such *et al.* (2000a,b); Szymonski *et al.* (2001)]. The formula (7.2) can be rewritten in the form

$$\Delta_0(J(t)) = \frac{bJ(t)}{a + J(t)}.$$

Using the potassium atom desorption signal reproduced in Fig. 7.7(a), it was possible to calculate the corresponding surface recombination coefficient shown in Fig. 7.7(b). On the other hand, NC-AFM images of the surface irradiated at the same temperature [Such *et al.* (2000a,b)] were used for calculating the number of edge atoms present on the surface after consecutive irradiations with electrons. The results are shown as points in Fig. 7.7(b). The density of the edge atoms is given in percentage of the total surface atomic density. The surface recombination coefficient, however, is calculated in arbitrary units and in order to be compared with the edge atom density it must be scaled accordingly. It is visible that the points representing measured density of edges are in perfect agreement with the scaled curve of surface recombination coefficient.

Chapter 8

Surface Patterning with Ions

Surface patterning by ion bombardment is considered as a convenient tool for manufacturing large area patterned templates for various applications. In all these applications it is highly desirable to create surface patterns with controlled length scales by properly adjusting the ion beam parameters such as the beam angle of incidence, current density, the ion energy, and the total fluence. The materials most frequently studied are metals and semiconductors. Up to the time of writing ion beam modification of ionic insulators, such as halides, has been only scarcely investigated, but recent results suggest that it may become an interesting alternative technique to electron and photon stimulated desorption.

8.1 Ripple Formation by Ion Bombardment

On metal and semiconductor surfaces it appears that nanoscale dots are created at close to normal beam incidence [Facsko *et al.* (1999)], whereas *ripple-like structures* are formed at more oblique angles of incidence [Valbusa *et al.* (2002); Makeev *et al.* (2002)]. Although there is a general consensus about evolution of the ripple structure as being due to the competition between surface roughening introduced by ion sputtering and surface smoothing created by various diffusion related processes, details of such mechanisms are still under discussion. Depending on the particular experimental conditions, pure thermally assisted diffusion [Bradley and Harper (1988); Herring (1951); Mullins (1957)], viscous flow [Chason *et al.* (1994); Erlebacher *et al.* (1999)], ion-enhanced, or -inhibited diffusion [MacLaren *et al.* (1992); Rossnagel *et al.* (1982); Ditchfield and Seebauer (1999)], and preferential sputtering without net mass transfer [Ditchfield and Seebauer (1999); Makeev and Barabasi (1997)] are considered as important contributors to the effective surface smoothing.

On halide surfaces, the fabrication of periodic wire arrays by 4.5 keV Ar^+ bombardment of a $CaF_2(111)$ crystal at grazing incidence (4-6 degrees with respect to the surface plane) at room temperature was reported [Batzill *et al.* (2001)]. The morphology of the irradiated surfaces was investigated 'ex-situ' with an ambient AFM operating in contact mode. The results indicated that the calcium fluoride surface was sputtered with preferential emission of fluorine, thus the excess of calcium at the surface assembled into periodic wire-like stress domains. The postulated self-organization mechanism invokes minimization of the system free energy and the elongation of the wires along the beam direction due to angle dependent anisotropic sputtering of the wires. Such a mechanism is contrasted with continuum theories [Bradley and Harper (1988); Makeev *et al.* (2002)] (see below), which in that case were assumed not to play any substantial role.

Surface morphology evolution of KBr and RbI single crystals under initial stages of ion bombardment with 5 keV He^+ was investigated using a NC-AFM in UHV [Krok *et al.* (2004a)]. AFM images taken for RbI(001) surface after consecutive ion irradiations at the crystal temperature of about 95 °C are reproduced in Fig. 8.1. The erosion started by creation of randomly spread rectangular pits of monatomic depth on the surface [Fig. 8.1(b)]. As the ion fluence increased the erosion proceeded by growth and linking of the pits in the first upper-most layer similarly to the ESD process described in the previous chapter, but the second layer was eroded before the first layer had been completely removed [Fig. 8.1(c)]. The *multi-layer* mode of desorption is even more distinctly visible in Fig. 8.1(d) where the pits in the third layer together with the remaining part of the first one are observed. Similarly, for the KBr(001) surface it was found that the surface topography evolution reflected the multi-layer mode of desorption [Krok *et al.* (2004a)].

Nanostructuring and optical activation of randomly oriented LiF polycrystalline samples by 0.8 keV Ar ion sputtering at different ion fluences was also investigated [Mussi *et al.* (2006)]. The angle of incidence was fixed at 35° off-normal and the sample temperature was 80 °C (Fig. 8.2). Although the AFM images were taken only for three different ion doses, the authors concluded that the trend in ripple height and the mean periodicity seemed to follow the predictions of a so called 'continuum theory' and the scaling law predicted by the MCB theory (see below).

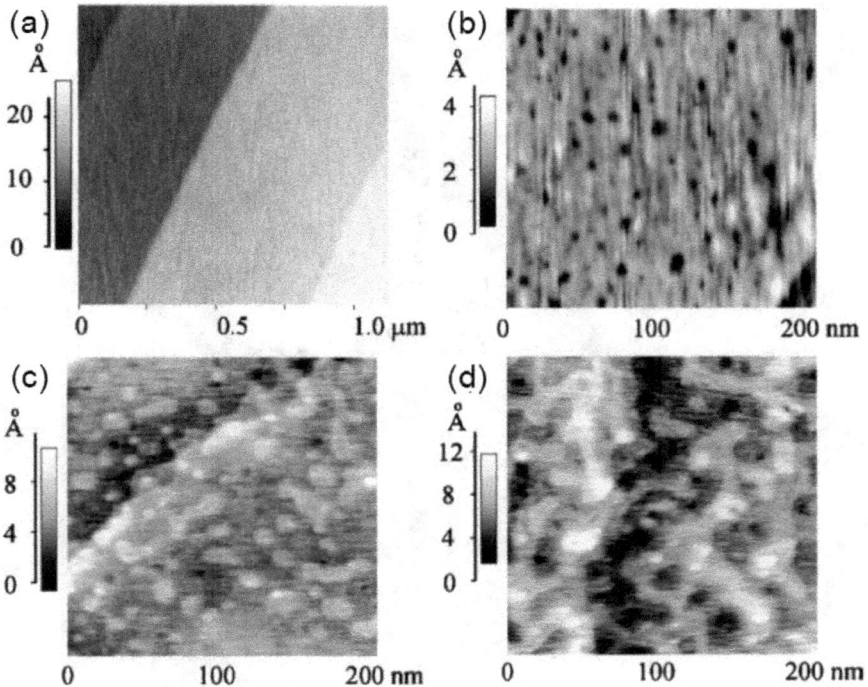

Fig. 8.1 NC-AFM images of cleaved RbI(001) surface: (a) as cleaved, (b) irradiated with an ion fluence of 0.43×10^{13} ions/cm^2, (c) irradiated with an ion fluence of 1.27×10^{13} ions/cm^2, (d) irradiated with an ion fluence of 2.62×10^{13} ions/cm^2. The irradiation was performed with 5 keV He$^+$ ion beam with an angle of incidence of 60° off normal. The sample temperature during the irradiation was 95 °C. Reprinted from [Krok *et al.* (2004a)], Copyright 2004, with permission from Elsevier.

8.1.1 *Linear continuum theory for ripple formation*

The *linear continuum theory* (LCT), initially outlined by Bradley and Harper [Bradley and Harper (1988)], has been commonly used to explain the process of ion-induced ripple formation in amorphous samples. The basic assumptions of the continuum theory imply that the local erosion yield is continuously and smoothly varying with the local surface morphology and the erosion is faster in depressions than at elevations due to a spatial extension of the ion-induced collision cascades [Sigmund (1973)]. This ballistic sputtering process causing surface coarsening is balanced by a diffusion-type surface smoothing processes of thermal nature [Bradley and Harper (1988)]. Since the linear theory could account for a limited

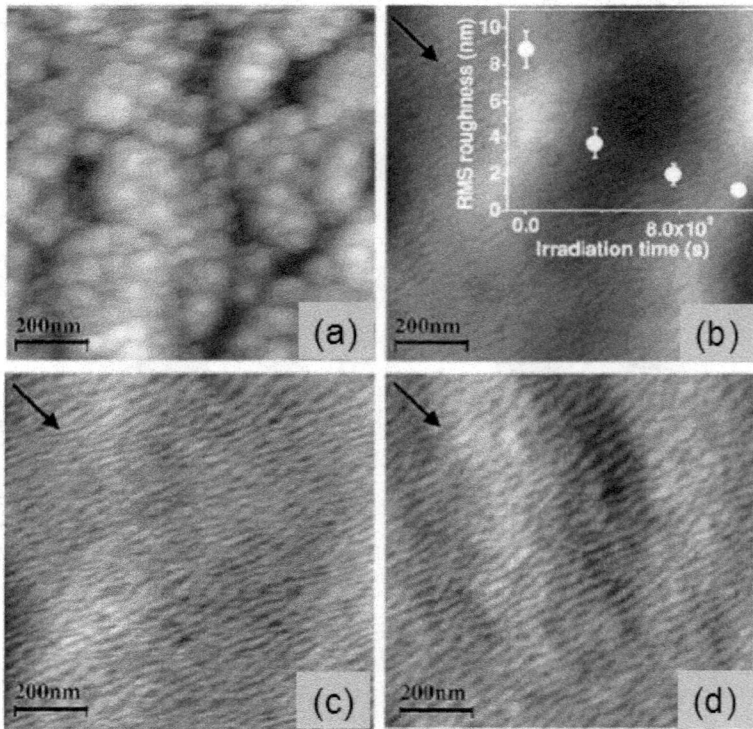

Fig. 8.2   AFM images (1 μm) of the surface of four LiF initially identical samples: (a) not sputtered (z dynamic range - 51 nm), (b) irradiated for 1 h, (c) 2 h and (d) 3 h with 800 eV Ar$^+$ ions at an angle of about 35°. The black arrow indicates the direction of the projected ion beam on the crystal surface. The graph reported in the inset shows the mean RMS roughness as a function of the irradiation time. Reused with permission from [Mussi *et al.* (2006)]. Copyright 2006, American Institute of Physics.

number of experimental findings only, several attempts were made to introduce additional effects and corresponding terms into the continuum theory description. In particular, Makeev, Cuerno, and Barabasi (MCB) generalized the Bradley and Harper's second-order linear equation by adding nonlinear and fourth-order terms to address the inadequacies of LCT [Makeev *et al.* (2002); Makeev and Barabasi (1997)]. The generalized theory introduced an ion-induced smoothing mechanism via preferential sputtering without mass movement on the surface. The ion-induced smoothing becomes important when the thermal diffusion is negligible, and, therefore, makes the ripple wavelength at low temperatures independent on the ion flux. In addition, this theory has many persuasive trends including a linear increase of the ripple wavelength with ion energy and a stabilization of the ripple amplitude for a prolonged ion bombardment. Some experiments, however, appear to indicate the flux independence of the wavelength over a certain range of fluxes [MacLaren *et al.* (1992); Rossnagel *et al.* (1982); Ditchfield and Seebauer (1999)] and that, for certain ion substrate combinations, there exists a minimum incidence angle (the angle between the ion beam and the surface normal) necessary for the formation of a ripple structure. Such findings question the generality of the MCB theory even for amorphous materials (or amorphizing due to ion bombardment).

8.1.2 *Beyond the continuum theory*

For single crystals, there are at lest two noticeable reports which could not be explained by the continuum theory. One work offered an atomic-scale description of the very shallow ripple formation (*nanogrooves*) upon grazing incidence ion bombardment of a single crystal Cu(001) surface [van Dijken *et al.* (1999)]. More recently, an atomic scale model for formation of ripples under grazing incidence Ar^+ bombardment of a Pt(111) surface was also considered [Hansen *et al.* (2006)]. One of the main experimental findings of that work indicates a striking difference between an *athermal* (200-470 K) and a *thermal* (470-700 K) pattern formation regime. It was stressed that the two temperature regimes are distinct not only by their different ripple wavelength dependence on the temperature, but also by the different pattern appearance: from ripple *V-shaped valleys* to ripple ridge one (*U-shaped valleys*). Furthermore, at the bottom of the U-shaped valleys (high temperatures) some compact vacancy islands are seen bound by the steep walls oriented along the ion beam direction, whereas only narrow vacancy grooves are seen at the bottom of V-shaped valleys (low temperatures).

The authors postulated that large differences in step edge versus terrace damage are crucial for the ripple formation under grazing incidence. In the low temperature (athermal) regime, initially the vacancy grooves are formed along the beam direction due to rapid motion of the irradiated step edges. Subsequent interaction of the neighboring grooves through repulsion of adatoms and the depletion effect prevent formation of the vacancy grooves on narrow adatom pads separating the neighboring grooves. At elevated temperatures (the thermal regime) alignment of vacancy structures is achieved by the preferential coalescence of compact vacancy clusters along the beam direction as a consequence of damage effects caused by the planar subsurface channeling. A large step edge barrier for vacancies to heal at ascending steps causes preferential production of new vacancy structures at the bottom of the vacancy grooves, transforming the initial pattern into well developed U-shaped ripple structure with increased roughness.

8.2 A Case Study: Ion Beam Modifications of KBr Surfaces

A comprehensive study of surface patterning processes induced by Ar ion bombardment has been recently reported for single crystal KI(001), RbI(001) and KBr(001) [Saeed *et al.* (2008)]. It is noteworthy to stress that the investigations were done in an UHV environment at every stage of the experiment, i.e. without exposing the samples to air for AFM imaging. Since alkali halide surfaces are very sensitive to moisture, they could easily be modified by water adsorption at improper vacuum conditions. This is especially important for irradiated samples containing a large surface density of structural and electronic defects [Yamada and Miura (1998)]. Surprisingly, it was found that the temperature dependence of the ion-induced KBr(001) surface morphology is quite similar to the Pt(111) case, despite large differences in properties of the two crystals. The two temperature regimes previously discussed [Hansen *et al.* (2006)] were clearly seen for KBr too, including the U-valley formation at elevated temperatures, each one covered at the bottom with rectangular vacancy islands bound in the step edges along the low index direction (Fig. 8.3).

In KBr the change between athermal and thermal regime of ripple formation seems to occur at the temperature corresponding to the onset of the ground state F-center mobility (above 500 K) [Salminen *et al.* (1996)]. At this temperature the halogen vacancies (F-centers) could migrate thermally to the surface and coalesce into larger vacancy agglomerates at the flat areas on the surface. The vacancy island equilibrium shape is rectangular with edges along the low index direction $\langle 100 \rangle$ and $\langle 010 \rangle$. In a previous work the same group reported on the step edge barrier for alkali halide molecule migration over the descending steps [Goryl *et al.* (2007)]. This barrier is responsible for pyramid-like shape of the alkali halide islands grown epitaxially by molecular beam epitaxy. The vacancy islands covering the flat bottom areas of the U-shape valleys have the inverted pyramid shape, confirming the existence of such a step edge barrier.

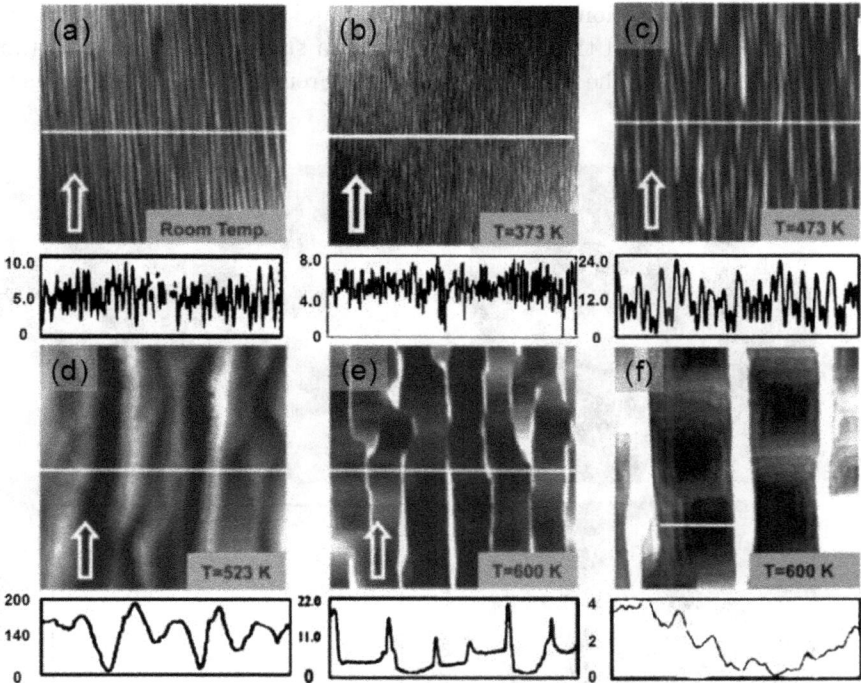

Fig. 8.3   KBr(001) surface modified with 5×10^{17} 5 keV Ar^+ ions/cm^2 striking at 75° off-normal. A set of AFM images (2 μm) obtained for samples exposed to the ion beam at (a) room temperature, (b) 100 °C, (c) 200 °C, (d) 250 °C, and (e) 325 °C. (f) Zoomed area (1 μm) of image (e). Cross-cut profiles (in nm) along the white lines on the images are shown below. Reprinted from [Saeed *et al.* (2008)], Copyright 2008, with permission from Elsevier.

Furthermore, the equilibrium surface morphology (for high fluences) critically depends on the angle of incidence, as shown in Fig. 8.4. Only for oblique angles (larger than 60° off normal) the ripples are very regular and nearly parallel to the projection of the incident beam. Therefore, the apparent similarity in ripple formation mechanism between platinum and potassium bromide single crystals would not be surprising. However, within the investigated energy range the ion sputtering processes of alkali halides are dominated by the electronic processes rather than the ballistic collisions [Krok *et al.* (2004a); Szymonski (1982)] despite a substantial nuclear stopping of keV heavy ions. This is due to a high probability of the valence excitations in those materials and a very high efficiency of radiation-less decay of those excitations with a subsequent transfer of the released energy into the lattice constituents (Frenkel pairs, direct halogen emission, etc.).

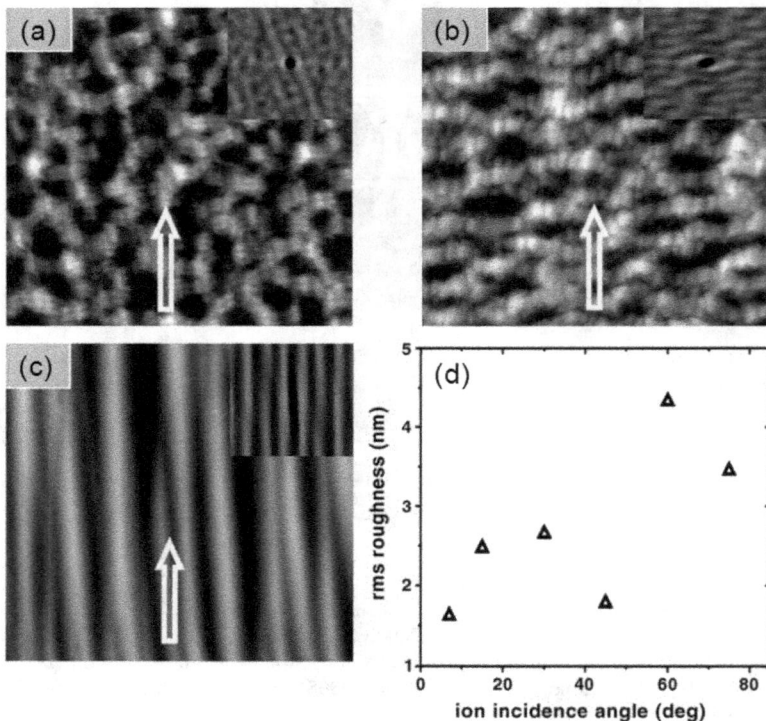

Fig. 8.4   Set of AFM images (1 μm) of a KBr(100) surface modified by Ar^+ irradiation for various incidence angles with respect to the surface normal: (a) 15°, (b) 30°, and (c) 75°. The insets in the upper-right corners show the 2D self-correlation function and the arrows indicate the ion beam direction. (d) Dependence of the RMS surface roughness on the angle of incidence. Reprinted from [Saeed *et al.* (2008)], Copyright 2008, with permission from Elsevier.

The surface morphology development was monitored step-by-step by increasing the ion bombardment time, i.e. by increasing the total irradiation fluence as shown in Fig. 8.5. It is seen that the solid-vacuum interface of the non-bombarded crystal consists of large atomically flat terraces with typical sizes exceeding 1 μm, separated by single atomic steps. At an ion fluence of about 10^{15} ions/cm^2 [Fig. 8.5(b)] clear corrugations are seen corresponding to vacancy clusters (black depressions) and irregular islands of adatoms (light elevated patches). At this stage no clear correlation of the surface pattern and the incident beam direction could be seen. A different situation is observed if the fluence increases by one order of magnitude [Fig. 8.5(c)]. In such a case a distinct ripple

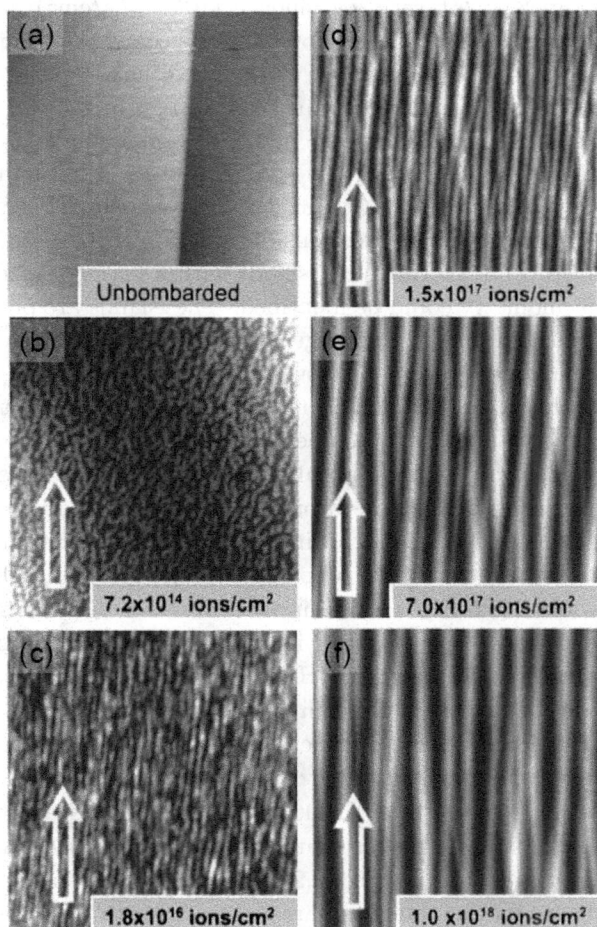

Fig. 8.5   Set of AFM images (1 μm) of a KBr(001) surface modified with various fluences of a 4.0 keV Ar$^+$ beam at $\theta_{inc} = 75°$ off-normal (b) - (f). (a) AFM image (2 μm) of the unbombarded surface. The surface projection of the ion beam direction is marked by arrows. Reprinted from [Saeed *et al.* (2008)], Copyright 2008, with permission from Elsevier.

structure is developed, although the ripples are not quite homogeneous. Further increase of the fluence results in formation of very regular pattern of ripples with increasing wavelength as seen in Fig. 8.5(d-f). The elongation of the ripples coincides perfectly with the ion beam projection on the irradiated surface.

A detailed inspection of the initial stages of the ripple formation process is shown in Fig. 8.6. At very low fluence (10^{13} ions/cm^2) a network of rectangular pits is formed on the bombarded surface [Fig. 8.6(a)], which resembles very much the appearance of the alkali halide surface subjected to a very low fluence of 1 keV electrons (Sec. 7.2). A further increase of the fluence results in 2D expansion of the pits with connecting and overlapping the neighboring features [Fig. 8.6(b)]. At this stage a clear indication of the groove-like morphology is seen with the grooves being elongated in one, or two main crystallographic directions of the (001) surface. In addition to pits and grooves, some elevated 2D islands with a perfect crystallographic order and stoichiometry could be identified. The islands are present on the irradiated areas close to the step edges. At a fluence approaching roughly 1×10^{15} ions/cm^2 the network of grooves and adatom islands is getting well ordered along the beam direction eventually transforming into the periodic ripple structure at about 10^{17} ions per cm^2. The exact mechanism of this transition is yet to be established, although the KBr data are consistent with the main ideas behind the atomic scale model of ripple formation proposed in [Hansen *et al.* (2006)]. Firstly, the erosion of the surface occurs predominantly by removal of the step edge atoms. In alkali halides, a lack of fast electronic relaxation results in an efficient energy transfer of local electronic excitations caused by ion irradiation into a production of well separated interstitial-vacancy pairs (H + F centers). Surface recombination of defects leaves behind pairs of alkali and halogen vacancies at the atomic step edges (Sec. 6.2). Some vacancies are also seen between the steps [Fig. 8.6(a)], possibly aggregating at the surface point defects created by ballistic collisions [Hansen *et al.* (2006)]. The 2D vacancy aggregates form a characteristic monolayer deep rectangular pits, well known from electron stimulated desorption studies (Sec. 7.2). Under a normal incidence there is no physical reason for any preferred directions of growth for the erosion pits. At an oblique incidence, however, the primary local excitations (electron-hole pairs and excitons) are produced along the incident ion track. Since the range for diffusion of the excitations is limited, the step erosion (of the atomic step edges and/or the edges of the pits) should proceed with preference for the incident beam direction.

Fig. 8.6 AFM images of the initial stages of KBr(001) surface modification with a 4 keV Ar ion beam at the angle of incidence 75° off-normal. The sample temperature during the irradiation was 300 K. The data obtained for the fluence 3×10^{13} ions/cm^2 are shown in (a), (c) and (d), whereas data for 1×10^{14} ions/cm^2 are in (b). Reprinted from [Saeed *et al.* (2008)], Copyright 2008, with permission from Elsevier.

Since the electronic processes of alkali halide erosion known so far could not produce adparticles with equal amount of K and Br atoms [Szymonski (1993); Szymonski *et al.* (2006); Hess *et al.* (2005)], it can be assumed that the 2D islands are created as a result of ion-induced ballistic collisions, i.e. due to a nuclear stopping of the projectile. The scenario of adatom production by the *interplanar channeling* was proposed in the case of ripple formation in a single crystal Pt [Hansen *et al.* (2006)]. The planar channeling of the incident ion entering the interplane opening at the surface atomic edge (and/or the pit edge) could result in surface atom/molecule expulsion at the final end of the channeled ion track. As already mentioned above, alkali halide molecules deposited in the process of homoepitaxy from the molecular beam have a tendency to agglomerate into 2D islands rather than to descent into the lower layers to fill vacancies. Therefore, the adatoms, once produced by channeled ions, could likely account for the 2D islands seen on AFM images in Fig. 8.6. A second possibility which might be considered for ion bombarded alkali halides invokes thermal spike processes induced by the incident ions due to a high density of the deposited ion energy and slow dissipation of this energy out of the collision cascade volume. The spike mechanism was previously considered as a likely candidate for explaining the intermediate peak in the energy distributions of sputtered atoms [Szymonski (1982)] but islands produced as a result of re-crystallization of the spike volume should appear as 3D crystallites rather than perfectly 2D islands.

Although the results reproduced in Figs. 8.3-8.6 were obtained for a single crystal of KBr, they are not specific to this particular alkali halide. Other alkali halides, such as KI and RbI are also developing the ripple structure while subjected to low keV Ar bombardment [Saeed *et al.* (2008)]. Finally, ion irradiated alkali halides retain their crystallinity on the surface even for a prolonged bombardment at room temperature. This is due to high efficiency of the electronic processes, including defect diffusion and recombination of vacancies with the structural defects such as step edges, kinks and corners, preserving a high degree of the surface crystallographic order even upon the large fluence irradiation.

Chapter 9

Metal Deposition on Insulating Surfaces

Metal adsorbates on insulating surfaces are of great interest for applications ranging from catalysis to microelectronics, as well as from a fundamental point of view. Metal nanoclusters, in particular, may reveal chemical and physical properties that are quite different from their atomic and bulk material counterparts. For instance, gold nanoclusters deposited on oxide surfaces are highly reactive with various gases even at low temperature, in contrast to surfaces of bulk gold crystals, which are not catalytically active [Hammer and Norskov (1995)]. Palladium clusters on aluminum oxide are catalysts for cleaner methane combustion, where they help to reduce emissions of nitrogen oxide. This chapter presents several SPM investigations on metal nanostructures grown on insulating surfaces. Following the division operated in other chapters, we will first address metals on flat and structured alkali halide surfaces, and then focus on insulating and mixed conducting oxide substrates. Metals on thin insulating films and computer simulations of SPM imaging of metal clusters on insulating surfaces are also discussed.

9.1 Metals on Halide Surfaces

As a general remark, we note that the growth of metal structures on alkali halide surfaces is considerably affected by thermal effects. For instance, substrates may start to evaporate before reaching a temperature that would otherwise activate epitaxial orientation of the nanoclusters. Thus, even if metal nanowires can be observed at high coverages, in most cases metal nanoclusters tend to remain well separated from one another.

9.1.1 *Metals on plain halide surfaces*

Several AFM studies have addressed the deposition of metal adsorbates on alkali halide surfaces, which usually results in the formation of nanoclusters rather than in ordered metal films. Self-assembly of iron and FePt on NaCl(001) was investigated using a variable-temperature UHV-AFM [Gai *et al.* (2002a,b)]. Fig. 9.1 shows iron nanoclusters grown in absence of a wetting layer, i.e. in the Volmer-Weber mode (Sec. 2.3). The clusters appear quite uniform in size. By increasing the iron coverage the clusters became larger and taller, although their lateral and vertical size distributions remained narrow. Such dispersion is remarkable, considering that the Volmer-Weber growth generally results in much broader size

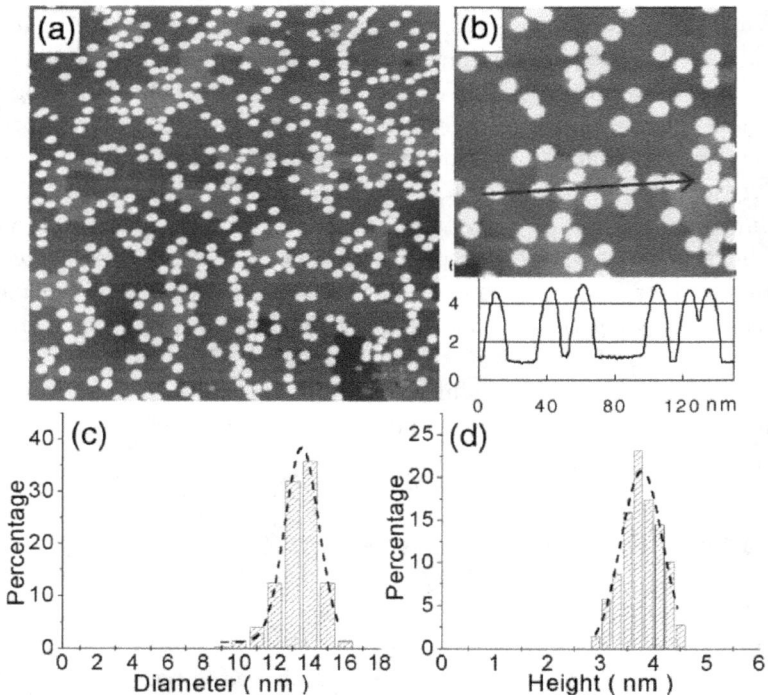

Fig. 9.1   (a) NC-AFM image (500 nm) of iron clusters grown on a NaCl(001) surface at a substrate temperature of 260 °C. (b) Close-up (200 nm) and line profile of the nanoclusters. (c) Diameter and (d) height distributions of the clusters shown in (a) and corresponding Gaussian fits. Reprinted with permission from [Gai *et al.* (2002b)]. Copyright (2002) by the American Physical Society.

distributions even for clusters formed by single elements [Moriarty (2001)]. The experimental observations were interpreted by a phenomenological model, in which a coverage-dependent optimal cluster size is defined by strain-mediated interactions among the clusters. Similar to pure iron, also FePt clusters were found to grow on the NaCl(001) surface without the formation of a wetting layer [Gai *et al.* (2005)]. By slightly miscutting the substrate the FePt nanoclusters could be spatially aligned, as shown in Fig. 9.2.

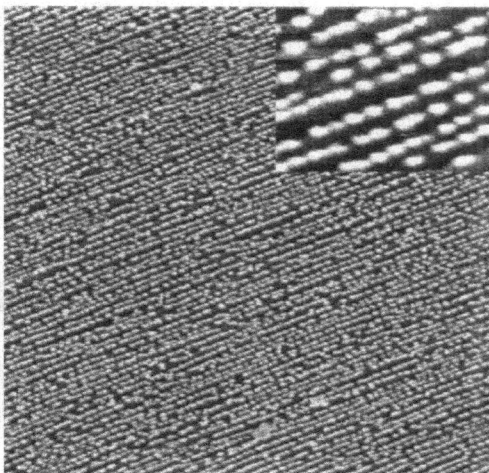

Fig. 9.2 AFM image (1 μm) of spatially ordered FePt clusters grown on a stepped NaCl(001) surface (cleaved at 3°) with a nominal thickness of 1.3 ML at a substrate temperature of 310 °C. The inset picture is a close-up (110 nm) of the cluster chains. Reused with permission from [Gai *et al.* (2005)]. Copyright 2005, American Institute of Physics.

High-resolution NC-AFM images of gold nanoclusters decorating the cleavage steps of KBr, NaCl and KCl(001) were also reported [Barth and Henry (2004, 2006b)]. The lateral size and shape of the clusters could be resolved more accurately using the microscope in constant height mode (Fig. 9.3). In such a case, the contrast could be enhanced and the clusters appeared much sharper compared to images acquired with constant Δf. This result can be attributed to a reduced tip-surface convolution caused by a sharp nanoasperity at the tip apex. KPFM images of gold clusters on KCl(001) [Barth and Henry (2006a)] revealed uniform bright contrast, which varied, however, among different groups of clusters (Fig. 9.4). A

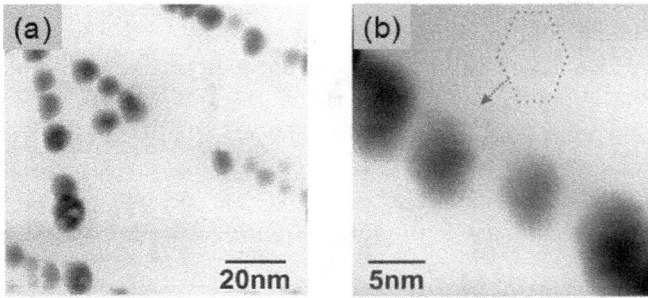

Fig. 9.3 (a) Gold nanoclusters on a NaCl(001) surface imaged in constant-height mode NC-AFM. The clusters are aligned at cleavage steps of the substrate. (b) Image revealing the hexagonal shape of some clusters. From [Barth and Henry (2006b)]. Reproduced with permission from IOP Publishing.

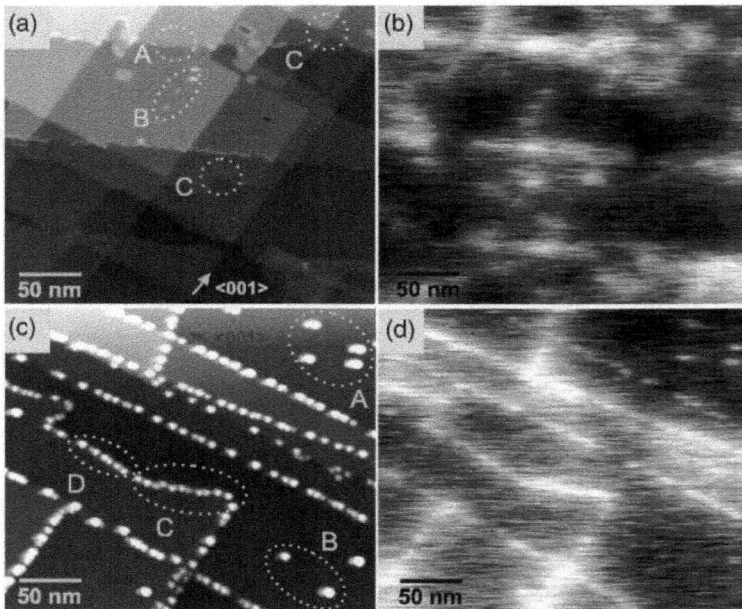

Fig. 9.4 (a) Topography and (b) potential images of a clean KCl(001) surface. The contrast is enhanced at kink sites (A), on KCl fragments (B) and at corner sites (C). (c) Topogaphy and (d) potential images of the KCl(001) surface after depositing 0.04 Au monolayers at room temperature. Some clusters on terraces (A) exhibit a brighter contrast than others (B). Similar behavior is recognized on rows of clusters (C vs. D). Reused with permission from [Barth and Henry (2006a)]. Copyright 2006, American Institute of Physics.

possible explanation is that the clusters got charged from the charges of the clean surface produced by cleavage (Sec. 2.2). The uniform contrast in each group is possibly due to a tunneling process of charges between clusters. Measurements after a charge injection into a cluster indeed revealed a possible exchange of charges in rows of nanoclusters.

9.1.2   *Metals on nanopatterned halide surfaces*

As shown in Sec. 2.2 iron nanoclusters have been used to demonstrate faceting of alkali halide surfaces. An early investigation reported the formation of iron nanowires on step edges of a NaCl(110) surface [Sugawara *et al.* (1997a)]. These wires had an estimated length of up to 10 μm and also revealed a strong in-plane magnetic anisotropy. On a similar surface, gold nanowires could also be observed, but only at high coverages [Kitahara *et al.* (2003a,b)]. Reducing the deposition time, only nanoclusters along the facets were formed.

The deposition of metal nanoclusters on substrates patterned with two-dimensional pits produced by electron- and photon-stimulated desorption (Chapter 7) has been also studied. In an early investigation of such phenomenon, gold decoration was used to image the edges of PSD-induced nanopits using a scanning electron microscope (Fig. 9.5). More recently, gold nanoclusters were deposited on a KBr(001) surface patterned with nanopits produced by ESD, and then imaged by NC-AFM [Goryl *et al.* (2007)]. As shown in Fig. 9.6, the nucleation of the clusters occurred preferentially on the upper side of the step edges. Thus, it is possible that the gold atoms landing inside the bottom of the pits escaped from them by means of thermally activated *upward diffusion* at the step edges.

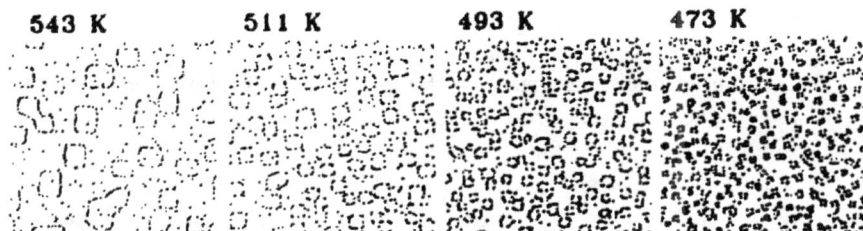

543 K   511 K   493 K   473 K

Fig. 9.5   Photon-stimulated two dimensional pits on a NaCl(001) surface observed by SEM at various crystal temperatures after 120 minutes of photon irradiation. Frame sizes: 1 μm. Reprinted with permission from [Hoeche *et al.* (1994)]. Copyright (1994) by the American Physical Society.

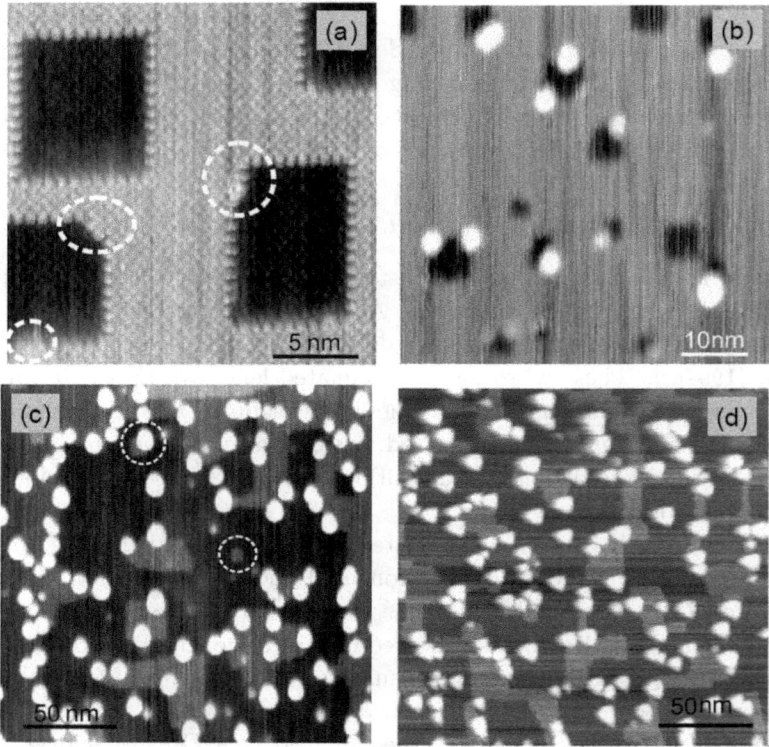

Fig. 9.6 (a) Atomically resolved NC-AFM image of nanoscale pits bound by [001] steps obtaining after exposing a KBr(001) surface to 1 keV electrons. (b) NC-AFM image of the same surface after deposition of 0.1 Au monolayers at room temperature. (c) Same as (b) with nanopatterns obtained by increased electron dose. (d) A NC-AFM image (200 nm) of a RbI(001) surface irradiated with UV light is shown for comparison. Reprinted with permission from [Goryl *et al.* (2007)]. Copyright (2007) by the American Physical Society.

Preferential gold nucleation sites at the edges of the nanopits may also be activated by the F centers produced in the electronic processes causing the patterning of the ionic surface.

Other NC-AFM measurements focused on tantalum atoms *codeposited* from an electron beam evaporator on a patterned KBr(001) surface [Mativetsky *et al.* (2006)]. In such a case, the pits surrounding the tantalum appeared much smaller than the other pits, and the pit density was nearly doubled compared to samples without metal adsorbates [Fig. 9.7(a)].

Tantalum nanoclusters decorating a KBr step, initially rounded and then modified in favor of ⟨001⟩ directions, are shown in Fig. 9.7(b). This result suggests that the clusters acted as nucleation sites for the growth of the nanopits.

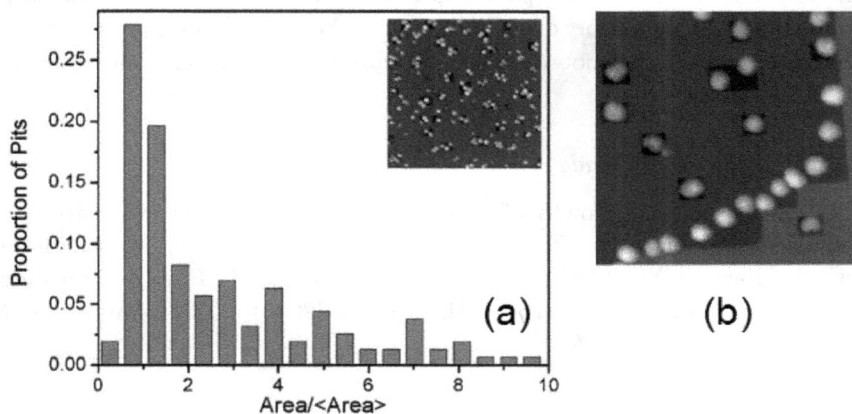

Fig. 9.7 (a) Pit size distribution on a KBr(001) surface with Ta clusters deposited during electron irradiation. The inset shows a 300 nm large area. (b) Ta nanoparticles inside pits and at receding KBr step. Reused with permission from [Mativetsky *et al.* (2006)]. Copyright 2006, American Institute of Physics.

9.2 Metals on Oxide Surfaces

The interactions between metals and insulating oxides are strongly dependent on the surface properties of the substrates. Surface stoichiometry, terminations and defects are the most important factors influencing the growth of metal adsorbates. The resulting structures have been mainly imaged by STM, after adequate treatments of the substrates.

9.2.1 *Metals on true insulating oxide surfaces*

9.2.1.1 *Aluminum oxide*

The interaction of metals with alumina surfaces is quite complex. Theoretical and experimental results have shown that the adsorption of reactive metals onto α-Al$_2$O$_3$(0001) is often accompanied by charge transfer from the metal adatoms to the substrate. In the adsorption of non-reactive metals, such as palladium, the charge transfer, however, occurs from the alumina surface to the metal [Fu and Wagner (2007)]. Scanning probe microscopy studies on these system are scarce, and obviously limited to AFM measurements. Investigations on Pd nanoclusters grown on Al_2O_3 showed some evidence of step-edge dominated growth [Pang *et al.* (2000)], similar to the behavior of palladium observed on thin alumina films (Sec. 9.3). This was not the case for copper nanoclusters addressed in the same study.

9.2.1.2 *Magnesium oxide*

The nucleation and growth of metals on MgO(001) surfaces is strongly influenced by surface defects. Fig. 9.8 shows the density of palladium islands grown on MgO(001) as determined from NC-AFM images in a wide range of substrate temperatures [Haas *et al.* (2000)]. The island density

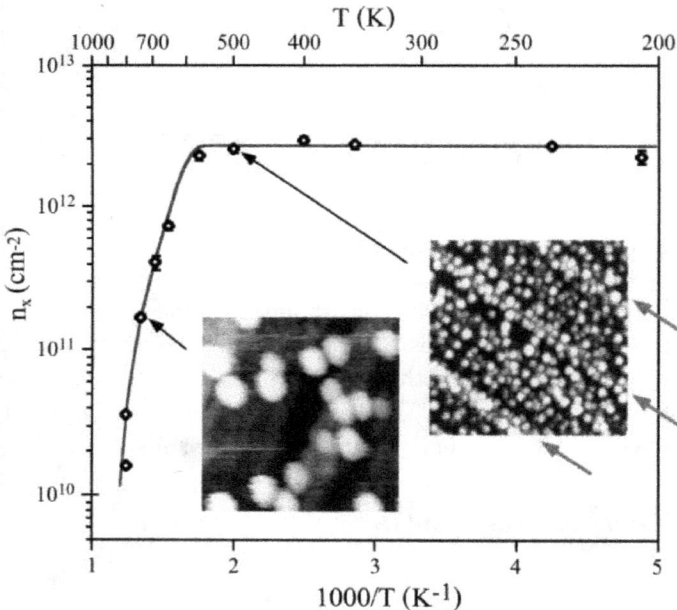

Fig. 9.8 Arrhenius representation of Pd island density on MgO(001) at 0.1 ML coverage. The solid line is the best fit obtained with a rate equation model. Two NC-AFM images (100 nm) are shown as insets. Three arrows point along steps in the 500 K micrograph. Reprinted with permission from [Haas *et al.* (2000)]. Copyright (2000) by the American Physical Society.

remained constant over a wide range of temperatures, indicating that the nucleation was governed by point defects with high trapping energy. The experimental results were interpreted within a rate equation model. As a result, energies of about 1.2 eV were estimated for adsorption and binding energies, slightly higher for defect trapping, and much lower (about 0.2 eV) for surface diffusion, in good agreement with independent *ab initio* calculations. Heterogeneous nucleation at point defects was found to be the dominating process also in the growth of gold clusters on Mg(001) [Hojrup-Hansen *et al.* (2004)].

9.2.2 *Metals on mixed conducting oxide surfaces*

Due to the mobility of both electronic and ionic defects in a mixed conducting oxide, the interaction between metal and substrate depends on both ionic and electronic conductivities. The electronic interaction is mainly determined by the electronic properties of the oxide, which tends to behave like a semiconductor. On the other side, the chemical interaction involves atom diffusion at the interface so that the transport of ionic defects in the oxide must be taken into account. In such a case, the oxide can be treated like an ionic solid.

9.2.2.1 *Titanium dioxide*

Metal atoms adsorbed on the $TiO_2(110)$-(1×1) surface are expected to occupy different locations. Metals with low electronegativity should bind to the surface via surface oxygens. In such a case the adatoms maximize their oxygen coordination numbers by accommodating on three-fold coordinated oxygens or binding to bridging oxygens (Sec. 1.2). These assumptions were confirmed by STM measurements on potassium, calcium and vanadium nanoclusters [Pang *et al.* (2005); Bikondoa *et al.* (2004); Agnoli *et al.* (2003)]. An example, referring to V clusters, is given in Fig. 9.9. Noble metals, on the other side, are expected to bind with surface Ti atoms. Gold clusters on top of the bright rows arising from surface five-fold coordinated Ti atoms were indeed observed by STM, suggesting that Au atoms nucleate on top of the Ti cations [Lai *et al.* (1998); Tong *et al.* (2005)]. Adsorption of platinum and palladium on five-fold coordinated Ti atoms was also documented [Dulub *et al.* (2000); Takakusagi *et al.* (2003b)]. In a recent study, enhanced adhesion of Au clusters has been reported when $TiO_2(110)$ surfaces are oxidized [Matthey *et al.* (2007)].

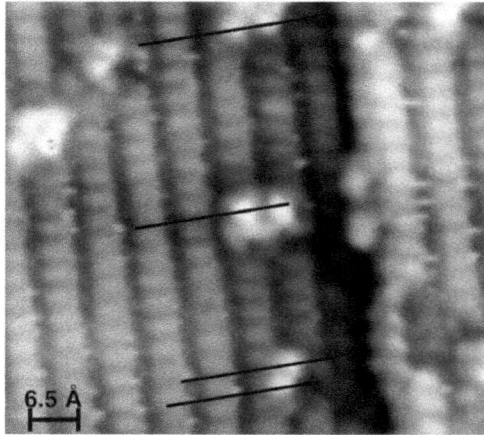

Fig. 9.9   STM image of 0.05 V monolayers on $TiO_2(110)$. The thin lines on the image indicate the registry of the V adatoms with respect to the maxima on the bright Ti rows. Reprinted from [Agnoli *et al.* (2003)], Copyright 2003, with permission from Elsevier.

9.2.2.2   *Strontium titanate*

The epitaxial growth of metal films on strontium titanate has been repeatedly investigated. Most metals grow on $SrTiO_3$ surfaces with the Volmer-Weber growth, although in few cases layer-by-layer growth was observed [Andersen and Møller (1991); Silly and Castell (2006)]. In any case the metal growth is strongly affected by the reconstructions of the substrate, which modify shape and orientation of the supported metal nanocrystals [Silly and Castell (2005); Silly *et al.* (2005)]. An example is given in Fig. 9.10, showing hut-shaped nanocrystals of palladium grown on the (2×1) reconstruction of $SrTiO_3(001)$, and hexagonal and truncated pyramid nanocrystals on a $c(4\times2)$ surface. These structures resulted from different surface energies of the two reconstructed substrates and from the dependence of the interface energies between metal and substrate on the substrate crystallography.

9.3   Metals on Thin Insulating Films

Thin films of insulators on conducting (or semiconducting) substrates are an interesting option for the growth of metal nanoclusters.

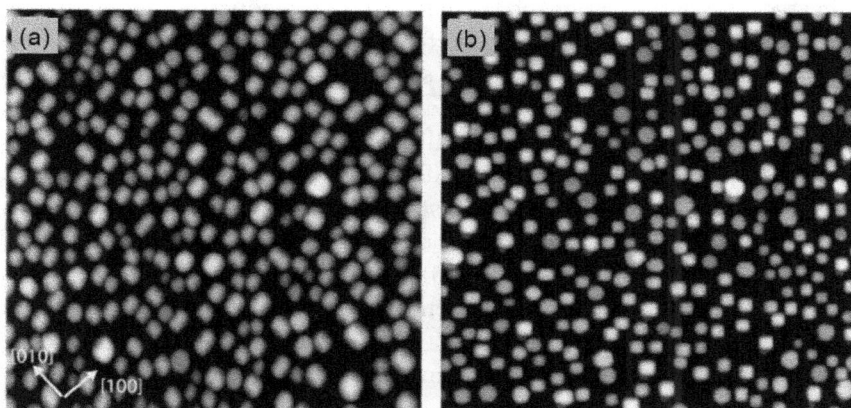

Fig. 9.10 (a) STM image (140 nm) of 2 Pd monolayers deposited on a $SrTiO_3(001)$-(2×1) surface at room temperature followed by annealing at 650 °C. (b) STM image (140 nm) of 1 Pd monolayer deposited onto $SrTiO_3(001)$-$c(4 \times 2)$ at 175 °C followed by 650 °C annealing. Reprinted with permission from [Silly *et al.* (2005)]. Copyright (2005) by the American Physical Society.

9.3.1 *Metals on halide films*

The growth of metals on halide thin films has been scarcely investigated. The formation of metal nanowires on the CaF_2/CaF stripes introduced in Sec. 5.2 was made possible by a specific preparation process. The substrate was first covered with ferrocene molecules, which self-assembled in the CaF trenches. As a second step the molecules were exposed to ultraviolet light in order to remove organic ligands, thus leaving an array of iron nanowires perfectly aligned on the surface [Lin *et al.* (2001)].

9.3.2 *Metals on oxide films*

In oxide film-based model systems the effect of the underlying metal support must be taken into account. If the thickness of the oxide layer is very small (usually less than 1 nm) electron tunneling between metal substrate and metal adsorbates occurs, and the support plays a major role. An example is given by thin alumina films grown on NiAl(110) [Kulawik *et al.* (2006)]. A STM study of gold adsorption on this system demonstrated indeed the formation of 1D gold chains with preferential orientations close to the [001] direction of the underneath NiAl substrate. Since the oxide film formed two domains, rotated of 48° against each other (Fig. 9.11), this result clearly indicated the participation of the substrate in binding Au to the oxide surface. The effects of the annealing temperature on cobalt, rhodium, and palladium nanoclusters grown on a similar substrate was systematically studied in a further investigation [Heemeier *et al.* (2003)]. Cobalt particles have the strongest interaction with alumina, and were indeed found to be the most stable. In all cases, the diffusion process was also found to be mediated by surface defects. Several studies focused on Co and Au clusters on the Al_2O_3/NiAl(001) substrate [Lin *et al.* (2005); Luo *et al.* (2006a,b); Lin *et al.* (2006); Sartale *et al.* (2006); Luo *et al.* (2008)]. As an example Fig. 9.12 shows chains of Co clusters a few hundreds of nanometers in length. It is interesting to note that these patterns remained intact even after flashing the sample to high temperatures, exposing it to atmosphere, or increasing the coverage up to coalescence.

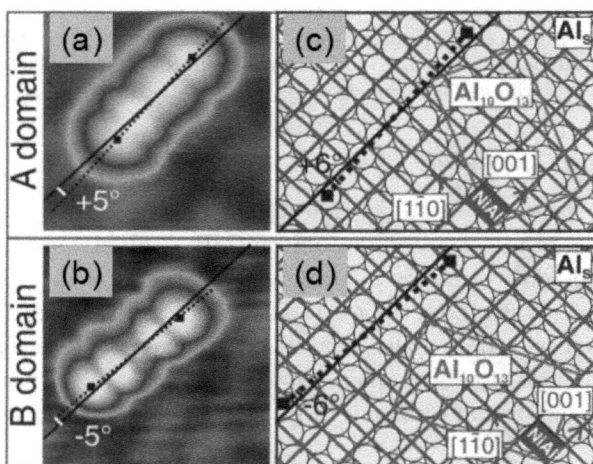

Fig. 9.11 (a, b) STM images of Au pentamers on an alumina film on NiAl(110). The angles between the chain (dotted lines) and the NiAl[001] direction (straight lines) are indicated. (c, d) Schemes showing the orientation of the hexagonal surface Al lattice with respect to the NiAl substrate. Reprinted with permission from [Kulawik *et al.* (2006)]. Copyright (2000) by the American Physical Society.

In the adsorption of metal atoms on MgO films, surface defects play an essential role, as in the case of bulk MgO. A recent STM investigation, combined with different surface science techniques, revealed that Au

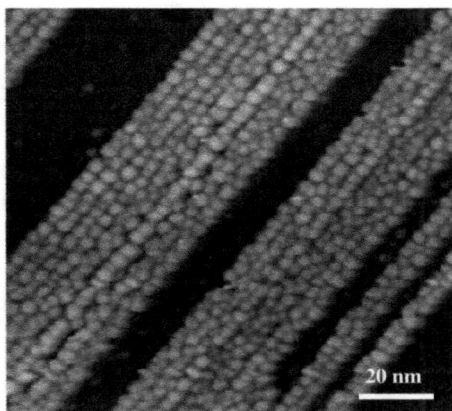

Fig. 9.12 STM image of Co nanoclusters formed at room temperature on crystalline Al_2O_3 grown on a NiAl(100) surface. The clusters aligned forming parallel chains. From [Luo *et al.* (2006a)]. Reproduced with permission from IOP Publishing.

clusters grown on a MgO film deposited on Ag(001) interact with F centers in a very specific way. The clusters adsorbed to oxygen vacancies were negatively charged while the gold clusters on regular terrace sites appeared neutral [Sterrer *et al.* (2006)]. Other examples of metal clusters on thin insulating films are given in Chapter 11, in the context of scanning tunneling spectroscopy.

9.4 Modeling AFM on Metal Clusters on Insulators

AFM imaging of metal clusters on insulating surfaces by NC-AFM was simulated in the case of gold clusters on KBr(001) [Pakarinen *et al.* (2006)]. In this study, tips covered with small amounts of Au or KBr were used. The variations in the tip-surface interaction over different cluster sites appeared negligible in both cases, and significant tip instabilities were observed, suggesting that atomic resolution may be very difficult to achieve in the experiments. Another study on palladium clusters on Mg(001) addressed the contrast formation in constant height imaging [Pakarinen *et al.* (2008)]. As shown in Fig. 9.13, high lateral contrast on the nanoclusters might be achieved with either a sharp nanotip or a significant short-range potential from the tip apex.

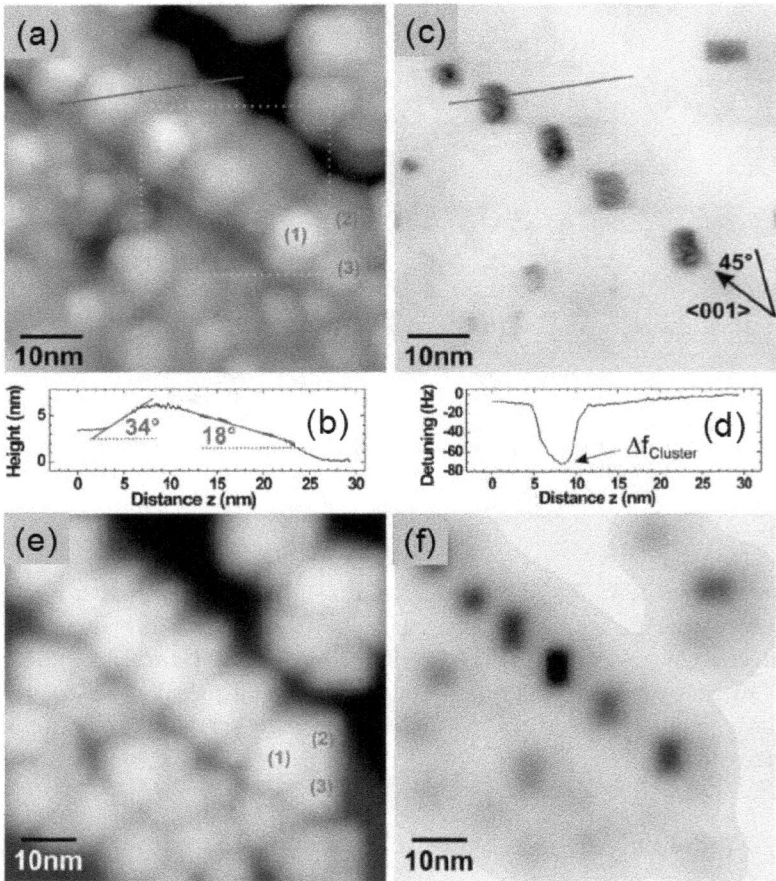

Fig. 9.13   (a, c) Experimental and (e, f) simulated images of Pd nanoclusters on MgO(001). Images on the left side are standard NC-AFM topographies, whereas images on the right were recorded or calculated in constant height mode. The highest clusters, aligned in a row, are attached at a [001] step of the surface. (b, d): Profiles taken at the positions of the gray lines in the experimental images. Reprinted with permission from [Pakarinen *et al.* (2008)]. Copyright 2008, American Institute of Physics.

Chapter 10

Organic Molecules on Insulating Surfaces

The interest in organic electronic devices, based on ordered films of organic molecules self-assembled on crystal surfaces, is growing rapidly. The operation of these devices should be ultimately governed by the electronic properties of a single or a few molecules. In this framework, insulating surfaces are required to avoid coupling between molecular electrons and substrate. While STM is an established technique to analyze organic molecules on metal substrates, imaging of molecules on insulators by AFM is still in early stages. In this chapter, after introducing the families of organic molecules that have been resolved by SPM techniques on insulators, we discuss the self-assembly of fullerene molecules on alkali halides, and the application of nanopits and cleavage steps as traps for molecular islands and guidelines for molecular nanowires. Molecular structures on mixed conducting oxides and thin insulating films can be imaged with high resolution by STM, and are discussed in separate paragraphs. Theoretical simulations aiming to reproduce AFM imaging of organic molecules on insulating surfaces are addressed at the end of the chapter.

10.1 Chemical Structures of Organic Molecules

In this paragraph we introduce some of the organic molecules that have been resolved by SPM techniques on insulating surfaces. The chemical synthesis of these molecules goes beyond our goals and is not discussed.

10.1.1 *Fullerene molecules*

Fullerenes are a family of carbon allotropes shaped as hollow spheres, ellipsoids, tubes, or planes. Minute quantities of fullerenes can be found

115

in nature, hidden in soot and formed by lightning discharges in the atmosphere. The smallest and most common fullerene is C_{60} [Kroto *et al.* (1985)]. This molecule has a cage structure, which is formed by 12 pentagons and 20 hexagons resembling a soccerball [Fig. 10.1(a)].

10.1.2   *Porphyrin molecules*

Porphyrin molecules are based on four pyrrole rings (C_4H_4NH) joined by methyne bridges, as shown in Fig. 10.1(b) [Smith (1975)]. Many porphyrins occur in nature, such as in green leaves and in red blood cells. Porphyrin molecules easily combine with metal atoms, which can be accommodated in the central cavity of the molecules. The chemical structure of a Cu-TBPP molecule is shown in Fig. 10.1(c). In this molecule four di-tert-butylphenyl (TBP) substituents form legs, which are able to decouple the porphyrin ring from the substrate onto which the molecule is deposited.

10.1.3   *Phthalocyanine molecules*

Phthalocyanine (Pc) molecules have the alternated N-C ring structure shown in Fig. 10.1(d). These molecules have a fourfold symmetry and can host hydrogen and metal cations in their center by coordinate bonds with the nitrogen atoms in the four isoindoles (C_8H_7N). The central atom can carry additional ligands. Since most of the elements can coordinate to a Pc molecule, a variety of Pc complexes exist. Among them, subphthalocyanines (SubPc's) are large polar molecules, which differ in several ways from the usual fourfold symmetric Pc's [Claessens *et al.* (2002)]. The central metal atom is replaced by boron, which is sp_3-coordinated to a chlorine atom placed in apex position and to three, instead of four, isoindoline rings [Fig. 10.1(e)]. This gives rise to a strong molecular dipole moment of about 1 eÅ.

10.1.4   *Perylene molecules*

Perylene is a polycyclic aromatic hydrocarbon with the chemical formula C_2OH_{12}, occurring in nature as a brown solid. Its derivatives, especially PTCDA (perylene tetracarboxylic dianhydride) molecules [Fig. 10.1(f)], have long been considered an archetype for organic electronics, because of their extended π-system and their ability to grow in a well ordered fashion on a wide range of substrates [Forrest (1997)].

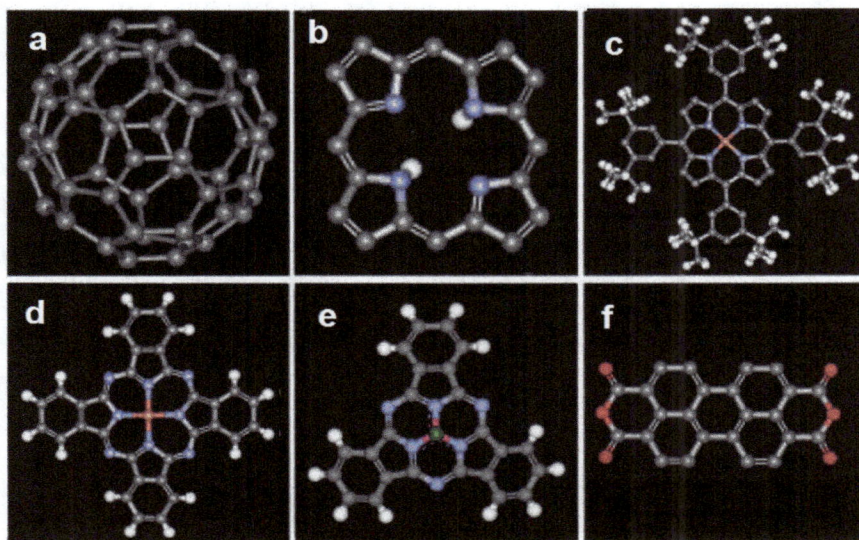

Fig. 10.1   Chemical structures (arbitrary scaled) of (a) C_{60}, (b) Porphyrin, (c) Cu-TBPP, (d) Cu-Pc, (e) SubPc, and (f) PTCDA molecules. Different colors correspond to different atomic species, according to the following scheme: grey = Carbon, white = Hydrogen, dark red = Oxygen, light red = Copper, green = Chlorine, light blue = Nitrogen, pink = Boron. Courtesy of Dr. Laurent Nony, Univ. Aix-Marseille.

10.2   Organic Molecules on Halide Surfaces

Self-assembly of organic molecules on alkali halide surfaces is strongly influenced by the molecular polarity. When apolar molecules such as C_{60} are deposited, the substrate has a scarce influence on the molecular arrangement. This is not the case for SubPc's or porphyrin molecules with functional groups, which are quite sensitive to the local electric fields on the substrate and tend to form regular nanowires driven by their dipole moments.

10.2.1   *Self-assembly of fullerene molecules*

Molecular resolution of C_{60} islands on insulators was first achieved in early studies on NaCl(001) [Lüthi *et al.* (1994)]. Hexagonal and triangular islands were found to nucleate predominately at step edges, but a simple epitaxial relation could not be determined. An unusual branched morphology of C_{60} islands was reported more recently on the KBr(001) surface [Burke *et al.* (2005)]. Molecular islands appeared at step edges (Fig. 10.2), as well as at screw dislocation centers. When the distances between steps were large, islands could be also observed on terraces [Fig. 10.2(d)]. Compact hexagonal crystallites were seen as well. The branched islands were often structured with an inner region and an outer rim corresponding to 1 and 2 C_{60} monolayers, respectively. From the distance between islands on flat terraces, a diffusion length of the molecules of several hundreds of nm could be estimated (at room temperature), which is consistent with previous studies by contact mode AFM [Yase *et al.* (1998)]. An alignment of one of the edges with one of the crystallographic directions of the substrate could be also recognized on the hexagonal islands at step edges, but not on the islands grown on terraces.

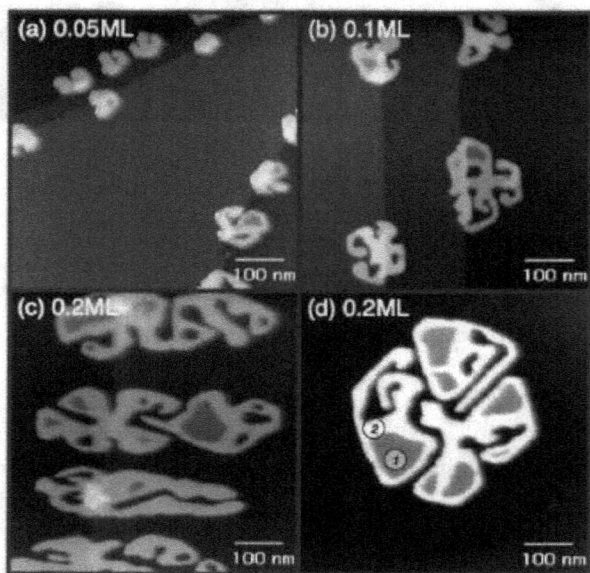

Fig. 10.2   NC-AFM images (600 nm) of (a) 0.05, (b) 0.1, (c) and (d) 0.2 C_{60} monolayers on KBr(001). Regions 1 and 2 are respectively 1 and 2 ML high. Reprinted with permission from [Burke *et al.* (2005)]. Copyright (2005) by the American Physical Society.

By acquiring images (not shown here) with atomic resolution on the KBr(001) substrate and, at the same time, molecular resolution on the C_{60} islands, a clear alignment of the high-symmetry direction with the $\langle 100 \rangle$ direction of KBr and a 2:3 ratio of C_{60} molecules to KBr conventional unit cells in this direction were determined. Though a unique supercell could not be estimated from the AFM images, an evaluation of the molecular overlayer energies of the possible $n \times 3$ supercells suggested that the 8×3 supercell is favored. Two different effective molecular heights were also observed (Fig. 10.3). The height difference between brighter and darker molecules corresponded to the height difference between two different disposals of a truncated icosahedron, respectively with a hexagonal ring and a pentagonal ring in contact with the substrate (see inset in Fig. 10.3). Height differences between C_{60} molecules were also reported in STM studies on noble metals [Altman and Colton (1992, 1993); Pai *et al.* (2004)].

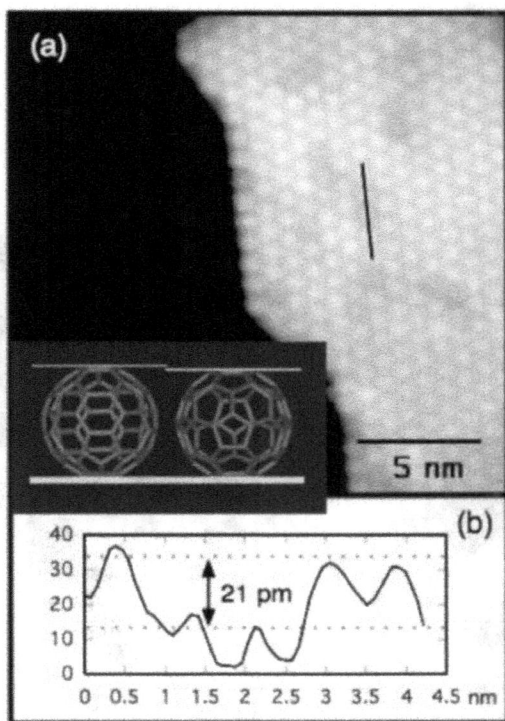

Fig. 10.3 (a) Molecular resolution of a single layer of C_{60} shows that some molecules are 'shorter'. (b) Cross section of the image above taken at the position marked by the black line, showing a height difference of 21 pm. The inset shows two orientations of the C_{60} molecules. Reprinted with permission from [Burke *et al.* (2005)]. Copyright (2005) by the American Physical Society.

Another investigation focused on the shape of C_{60} islands formed on both KBr and NaCl substrates [Burke *et al.* (2007)]. At room temperature, both compact and branched island morphologies were found to coexist. The branched morphology appeared only above a critical transition area (Fig. 10.4) and was attributed to a 'dewetting' process. Since intermolecular interactions are stronger than molecule-substrate interactions, the shapes of the islands looked similar, independently of the fact that the islands were grown on open terraces or driven by substrate steps. The branch widths and fractal dimension of the islands were almost the same on the two substrates. The only difference was the dependence of the transition area on the molecular coverage.

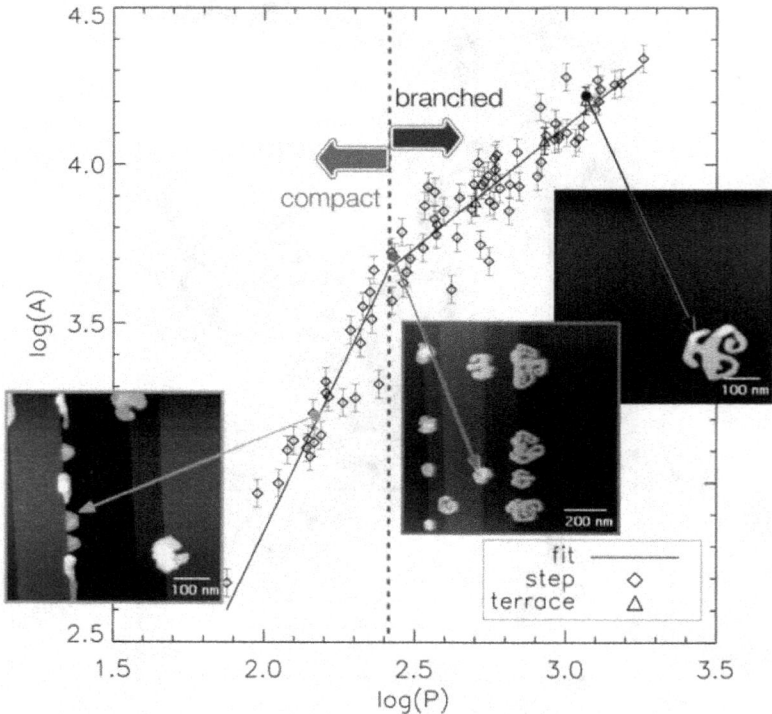

Fig. 10.4 A typical dimensionality plot [log(area) vs. log(perimeter)] for C_{60} islands grown on alkali halide surfaces. The slope changes from 2 to 0.9 at the transition between compact and branched shapes. Reprinted with permission from [Burke *et al.* (2007)]. Copyright (2007) by the American Physical Society.

10.2.2   *Nanoscale pits as molecular traps*

Rectangular pits produced by electron stimulated desorption on alkali halide surfaces (Chapter 7) have been successfully used to 'trap' organic molecules. This possibility was first demonstrated for PTCDA molecules forming flat assemblies inside KBr(001) pits less than 30 nm in width [Nony *et al.* (2004a)]. In another study, it was recognized that PTCDA molecules self-organize in a herringbone pattern inside the pits [Mativetsky *et al.* (2007)]. Here, most of the molecular islands appeared 3 or 4 ML high, with the first layer fully inside the pit and the top other layers protruding. The atomic structure of the KBr lattice could be also resolved simultaneously (Fig. 10.5). The growth of several layers of PTCDA molecules on KBr(001) has been also investigated [Kunstmann *et al.* (2005); Fendrich *et al.* (2007)]. While large areas of the substrate remained uncovered, some crystallites of different heights, from 2 up to 20-30 monolayers, could be recognized, even at low coverages. On the topmost layer of the crystallites, single PTCDA molecules could be resolved.

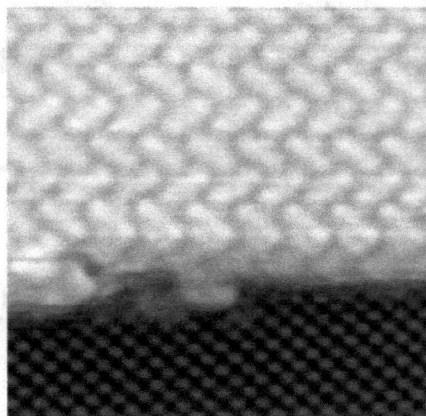

Fig. 10.5   NC-AFM image (10 nm) of a 3 ML PTCDA island grown inside a nanopit and the neighbouring KBr(001) lattice. From [Mativetsky *et al.* (2007)]. Reproduced with permission from IOP Publishing.

In another investigation, SubPc molecules were deposited onto an electron-irradiated KBr(001) surface [Nony *et al.* (2004b)]. In such case, the molecules self-assembled in the nanopits forming regular rows along the ⟨110⟩ direction, as shown in Fig. 10.6(a). The molecules ordered rather well in the center of the pits, but their disposal was somehow mismatched along the edges. Blurred bright areas, which can attributed to mobile molecules, and a few protruding molecules, are also visible on top of the organized layers. When attempting to image at decreased tip-sample distance, horizontal stripes appeared, implying that the tip started to drag material while scanning. Only pits below 15 nm in size were filled by the molecules, and no ordered aggregates were found on the flat terraces of the crystal. Although the conical shape of the molecules could not be resolved, a possible arrangement of the SubPc's was proposed, based on energetic considerations. Due to the ionic nature of the substrate, the electric field is significantly enhanced at the pit corners, suggesting that these locations trap the Cl apex of the first incoming SubPc molecules. The next molecules are expected to assemble in parallel rows at 45°, driven by their strong dipole-dipole interaction [Fig. 10.6(b)].

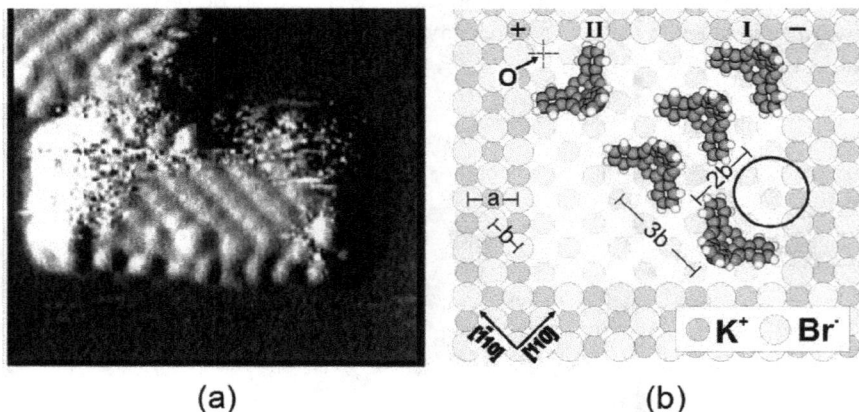

(a)   (b)

Fig. 10.6   (a) NC-AFM image (18 nm) of SubPc molecules trapped in an electron-irradiated KBr(001) surface. (b) Possible schematic arrangements of the molecules inside a pit consisting of $10 \times 10 \times 1$ missing ions. The Cl apex atom is strongly attracted by the electrostatic potential well near the corner site (position I). Position II is energetically unfavorable. Note that the molecular axis is slightly tilted. Reprinted with permission from [Nony et al. (2004b)]. Copyright 2004 American Chemical Society.

Molecular contrast was reported not only in NC-AFM topographies, but also in the damping signal [Pfeiffer et al. (2007)]. In the case of SubPc molecules on KBr, enhanced dissipation was observed when imaging specific edges of the molecular islands, probably due to the lower coordination of the molecules at such locations. A second layer of mobile molecules also exhibited increased damping.

10.2.3 *Molecular nanowires*

The formation of molecular wires on insulating surfaces is one of the most ambitious goals in molecular electronics. The vinylidene fluoride (VDF) molecules, formed by chains of strong dipole moments, were among the first species investigated on such purpose [Fukuma *et al.* (2002); Yamada *et al.* (2002)]. The arrangement of these molecules on KCl(001) revealed a characteristic honeycomb-like pattern on the first monolayer and rod-like structures on the following layers (Fig. 10.7), although a clear interpretation of these results could not be provided.

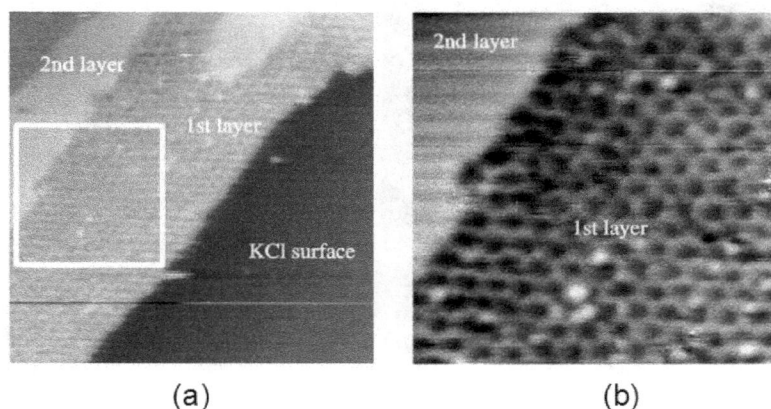

(a)   (b)

Fig. 10.7   NC-AFM images of a VDF film grown on KCl(001). Frame sizes: (a) 150 nm; (b) 60 nm. Reprinted from [Yamada *et al.* (2002)], Copyright 2002, with permission from Elsevier.

Cleavage step edges were soon recognized as natural templates for the growth of molecular wires. Step edges of KBr(001), for instance, were found to act as efficient traps for Cu-TBPP molecules [Fig. 10.8(a)], although the weak interaction between molecules and substrate resulted in the formation of irregular clusters rather than in ordered molecular arrays [Nony *et al.* (2004a)]. This is not the case on metal substrates, where ordered islands of Cu-TBPP could be observed at step edges [Loppacher *et al.* (2003)]. In another investigation porphyrin molecules were modified by linking their apolar core to a nitrogen atom [Zimmerli *et al.* (2005)]. This link induced a strong dipole moment, which increased the interaction with the substrate and prevented diffusion of the molecules. As a result, the molecules formed nanowires on straight step edges of the crystal, as shown in Fig. 10.8(b) in the case of rectangular pits on electron-irradiated KBr(001). At kink sites the molecular wires appeared interrupted, or did not form at all. More recently, rows of parallel nanowires formed by similar molecules were also observed on flat terraces of KBr(001) [Maier *et al.* (2008a)]. The internal structure of the molecules could be also partially resolved [Fig. 10.9(a)], suggesting that the porphyrins were π-π stacked and not lying flat on the surface [Fig. 10.9(b)].

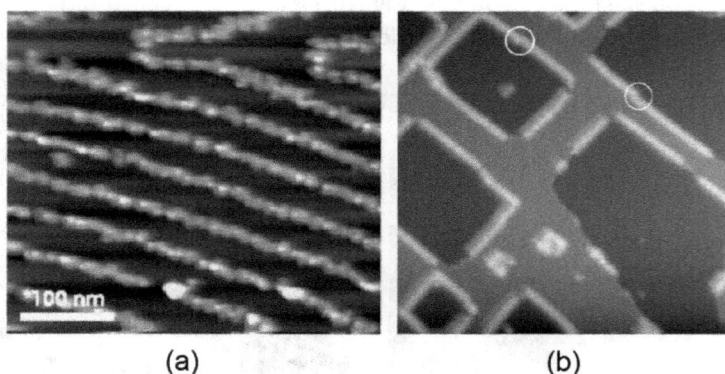

(a)   (b)

Fig. 10.8 (a) NC-AFM image of a KBr(001) surface after deposition of a submonolayer of Cu-TBPP molecules. Reprinted from [Nony *et al.* (2004a)]. (b) NC-AFM image (100 nm) of cyano-porphyrins on electron-irradiated KBr(001). Molecular wires are interrupted at kink sites. From [Zimmerli *et al.* (2005)]. Reproduced with permission from IOP Publishing.

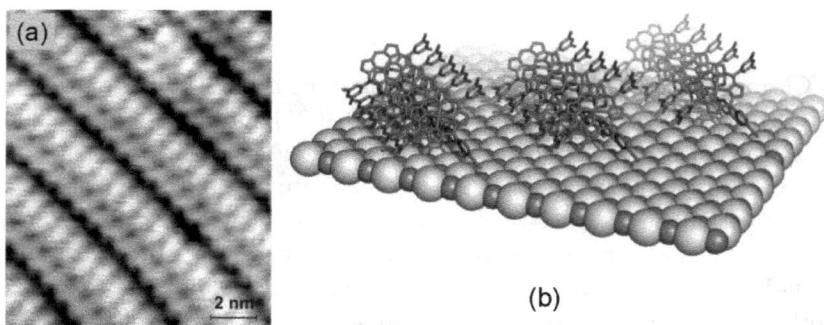

Fig. 10.9 (a) High-resolution NC-AFM image of a monolayer consisting of several parallel porphyrin wires on KBr(001). (b) A possible arrangement of the molecular wires. From [Maier *et al.* (2008a)]. Copyright Wiley-VCH Verlag GmbH & Co. KGaA. Reproduced with permission.

10.3   Organic Molecules on Oxide Surfaces

Only few SPM studies on organic molecules adsorbed on oxide surfaces have been reported. One of the earliest study of organic molecules on insulating oxides focused on a thin film of a phthalocyanine derivative grown on an alumina surface, which had previously been polished with a slight miscut [Hayashi *et al.* (1995)]. In such case, AFM images showed that the molecules were ordered by the miscut direction, rather than by the crystallographic orientation of the substrate.

A series of SPM investigations on acid molecules on titanium oxide was initiated by NC-AFM measurements on formic acid (HCOOH) deposited on a $TiO_2(110)$ surface [Fukui *et al.* (1997b)]. Fig. 10.10 refers to an STM study on formic acid with submonolayer coverage. Here, the number of bright spots increased with exposure, so that these spots could be associated to the adsorbates. Since the adsorbates were positioned over the bright rows on the substrate, and it is well known that formic acid reacts with exposed Ti, these rows were related to Ti sites. Other studies focused on acetic (CH_3COOH) and trifluoroacetic (CF_3COOH) acids on a similar surface [Sasahara *et al.* (2001a,b,c); Onishi *et al.* (2002); Sasahara *et al.* (2002, 2003)]. Figures 10.11(a) and (b) show both species of molecules arranged in a square symmetry. A mixed monolayer containing both molecules could be

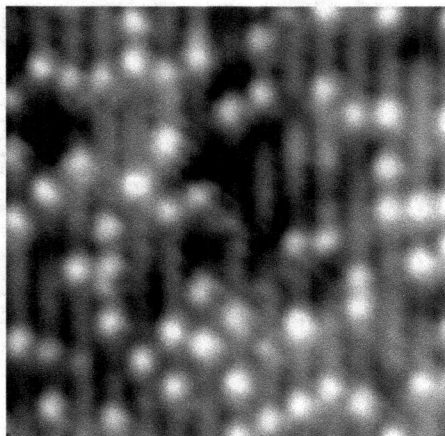

Fig. 10.10   STM image (10 nm) of a $TiO_2(110)$ surface exposed to formic acid vapor at room temperature. Reprinted from [Onishi *et al.* (1999)] with permision from from APEX/JJAP.

also recognized (not shown here). Fullerene molecules were also deposited on $TiO_2(110)$, and imaged by STM [Fukui and Sakai (2006)]. In such case, the peculiar structure of the substrate allowed the formation of molecular rows, such as those shown in Fig. 10.12.

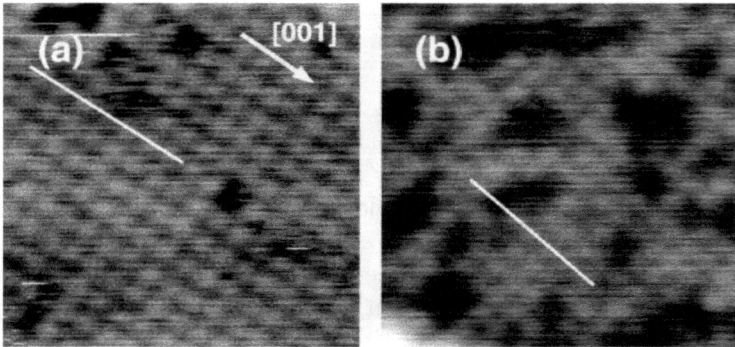

Fig. 10.11 NC-AFM images (9 nm) of (a) acetate and (b) trifluoroacetate monolayers on $TiO_2(110)$. Reprinted with permission from [Sasahara *et al.* (2001b)]. Copyright (2001) by the American Physical Society.

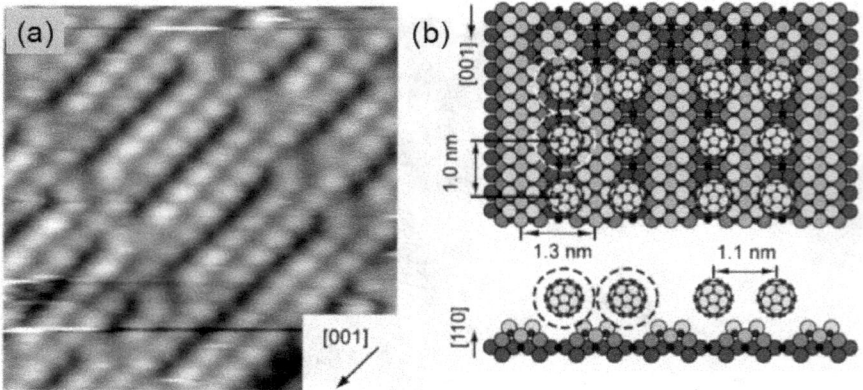

Fig. 10.12 (a) STM image (15 nm) of a C_{60} monolayer on a $TiO_2(110)$-1×2 surface. (b) Model of the system. The dotted circles represent van der Waals spheres of C_{60} molecules (1.05 nm). Reprinted with permission from [Fukui and Sakai (2006)]. Copyright 2006 American Chemical Society.

10.4 Organic Molecules on Thin Insulating Films

The use of ultrathin insulating films for the study of molecular layers is particularly promising. Here, the ordering of molecules strongly depends on the molecule-substrate interaction and, consequently, on the film thickness.

10.4.1 *Organic molecules on halide films*

A detailed STM investigation focused on porphyrin derivatives (CuOEP) deposited on NaCl films of different thickness, which had been previously grown on various metal substrates [Ramoino *et al.* (2006)]. The molecules formed extended domains with nearly hexagonal superstructures (Fig. 10.13), consistently with the point-on-line coincident growth mechanism introduced in Sec. 2.3. Each unit cell accommodated one molecule with the aromatic core parallel to the substrate. Furthermore, the molecules assembled in a sequential fashion, first on bare metal areas, then on the first and finally on the second NaCl layer. When scanning at low coverage, high surface mobility was observed. Such 'hierarchical' assembly can be attributed to the rapid decrease in the van der Waals interaction expected for an increased number of insulating layers and with an additional charge transfer to the metal.

The growth of PTCDA molecules on a Ag(111) surface partially covered by 1 or 2 monolayers of KBr was studied by NC-AFM [Loppacher *et al.* (2006)]. Different adsorption patterns were observed on the pure substrate and on the first and the second monolayer of the alkali halide film. These patterns could be interpreted using a modified Ising model, assuming that the adsorbate-substrate interaction dominates on the metal, the nearest and next-nearest neighbor interaction dominates on the 2 ML thick film, and all these interactions compete on 1 ML.

The self-organized CaF/CaF_2 stripes on stepped Si(111) surfaces introduced in Sec. 5.2 were used as adsorption patterns for organic molecules [Rauscher *et al.* (1999)]. Perylene derivatives were found to adsorb preferentially on CaF nanostripes, rather than on CaF_2. A possible explanation is that the π-electron system of the molecules interacts quite strongly with CaF, due to a good match of the molecular HOMO-LUMO gap and the band gap of the CaF layer. This property was exploited to fabricate large assemblies of parallel, equidistant stripes a few nanometers wide. The stronger electronic interaction, compared to the expected van der Waals interaction with the fully insulating CaF_2 stripes, facilitated the

Fig. 10.13 Zoom-in sequence of STM images of 1 ML CuOEP molecules on NaCl/Ag(111). The arrow refers to an antiphase boundary between two molecular domains with similar orientation. Reprinted from [Ramoino *et al.* (2006)], Copyright 2006, with permission from Elsevier.

selective adsorption of the molecules. The same thin-film template was also employed to guide the growth of ferrocene molecules (see also Sec. 9.3).

10.4.2 *Organic molecules on oxide films*

Similar to the case of bulk-truncated alumina substrates, ultrathin alumina film grown on $Ni_3Al(110)$ seem not to have a significant template effect on organic molecules. This was shown by STM investigations on CuPc molecules, which nevertheless distinguished two different kinds of adsorbate configurations [Qiu *et al.* (2004)]. More recently, a detailed STM study on the adsorption of CuPc on an alumina film on $Ni_3Al(111)$ revealed again two different molecular orientation, rotated by 30° with respect to each other, but only at submonolayer coverages [Moors *et al.* (2008)]. An example is given in Fig. 10.14. With increasing coverage the strong intermolecular π-electron interactions compensated the weak template effect of the surface resulting in the growth of 3D molecular clusters without any visible ordering.

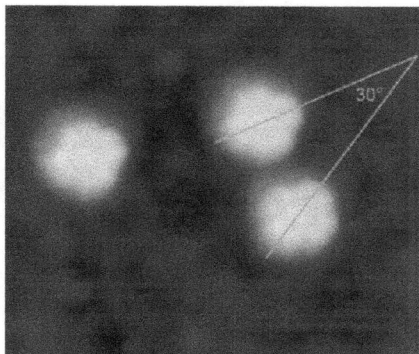

Fig. 10.14 STM image (13 nm) of three CuPc molecules adsorbed on $Al_2O_3/Ni_3Al(111)$. The different orientation of the single molecules with respect to each other is marked. Reprinted from [Moors *et al.* (2008)], Copyright 2008, with permission from Elsevier.

10.5 Modeling AFM on Organic Molecules on Insulators

Due to the complexity of the problem, only few theoretical studies have addressed AFM imaging of organic molecules. A simulation focusing on the interaction of monolayers of formic, acetic and trifluoroacetic acids on the $TiO_2(110)$ surface with a silicon dangling bond tip showed that the tip interacts more strongly with the substrate and the COO^- group than the adsorbed acid headgroups, so that the molecules should appear dark in NC-AFM images [Foster *et al.* (2005)]. In another study AFM images on small formic acid molecules on MgO(001) and on larger 3-propionic acid $(C_{52}H_{72}O_3)$ molecules on $TiO_2(110)$ were simulated using several tip materials and structures [Sushko *et al.* (2006)]. Again, the interaction of the molecules with the oxide surfaces appeared weak, suggesting that molecular adsorption should be favored by attaching special anchoring groups. In the latter study flat molecules could be identified by their shape, although simultaneous atomic resolution inside the molecule and on the substrate under the same imaging conditions appeared unfeasible. For instance, Fig. 10.15 shows a change in the pattern of a formate ion from a homogeneous ellipse to a pair of distinguishable blobs as a MgO tip advanced closer towards the molecule. The calculations also showed that an oxide tip with a positive potential can pick up a molecule from the surface if the tip is closer than about 0.5 nm from the surface. Therefore further approach to the surface can lead to tip contamination by adsorbed molecules.

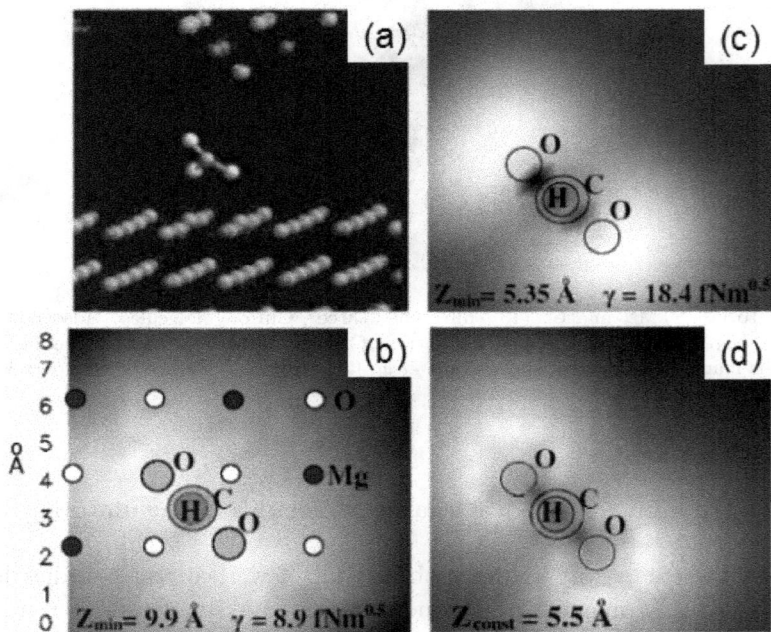

Fig. 10.15 Modeling AFM imaging of formate (HCOO$^-$) ions on MgO(001). (a) The adsorption geometry of the formate anion showing a soft bending mode induced by the tip. (b) and (c) Simulated images with decreasing tip-surface distance in constant Δf mode. (d) Constant height image with a sharp 10 nm tip scanning at a height of 0.55 nm. From [Sushko *et al.* (2006)]. Reproduced with permission from IOP Publishing.

Chapter 11

Scanning Probe Spectroscopy on Insulating Surfaces

Site-specific force spectroscopy is important to gain information on the contrast mechanisms in AFM imaging and, more generally, to understand interactions occurring at insulating surfaces. Once again, a vital role is played by theoretical models, whose predictions are essential to correctly interpret the experimental observations. On ultrathin insulating films grown on conducting surfaces, scanning tunneling spectroscopy (STS) is also feasible. Recent experiments have proven that STS can be also performed on single metal clusters and organic molecules deposited on the insulating films. Switching between different electronic states has been reported, which is very promising in the context of nanomanipulation. The goal of the present chapter is to summarize the main results obtained by force spectroscopy and STS on insulating surfaces and by STS on metal nanoclusters and organic molecules on insulating films.

11.1 Force Spectroscopy on Insulating Surfaces

The force on a probing tip as a function of displacement perpendicular to the sample was first measured on a Si(111)7×7 surface [Giessibl (1995)]. On this surface features like corner holes make relatively simple the separation between long-range and short-range contributions to the net force acting on the tip. More difficult is the case of insulating surfaces with the rock salt structure, that we are going to discuss at first.

11.1.1 *Alkali halide surfaces*

Force-distance curves can be better interpreted when they are acquired at low temperatures [Hoffmann *et al.* (2002, 2004); Lantz *et al.* (2006)].

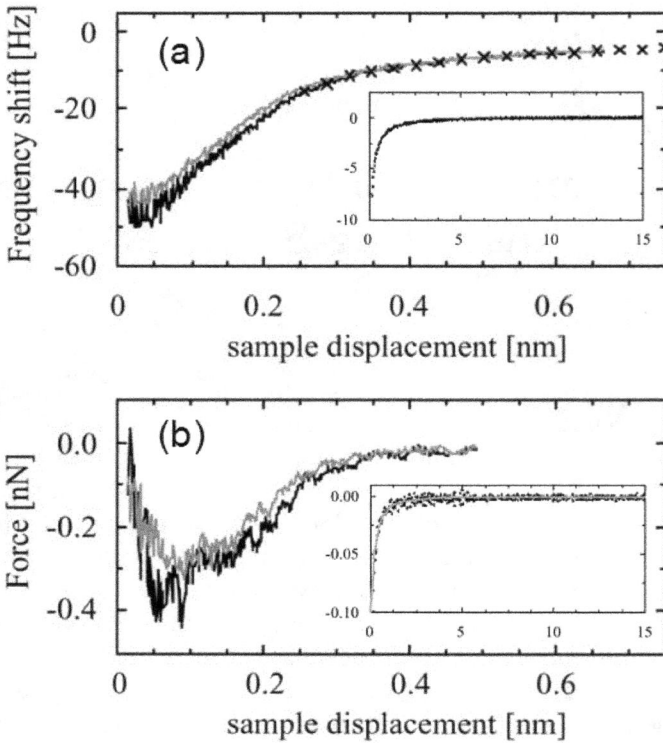

Fig. 11.1   Short-range frequency shift vs. distance curves on a KBr(001) surface. (b) Corresponding forces. Inset: Long-range components. Reprinted from [Hoffmann *et al.* (2002)], Copyright 2004, with permission from Elsevier.

Fig. 11.1(a) shows the frequency shift as a function of the tip-surface distance measured on a KBr(001) surface at 7 K. The corresponding force, determined with an iterative method (Sec. 3.1), is plotted in Fig. 11.1(b). In the range of distances where atomically resolved images could be acquired, force-distance curves were recorded with higher vertical resolution. The long-range components of the forces were fitted using the expression for the van der Waals force between a conical tip terminated by a spherical cap and a flat sample surface introduced in Sec. 3.1, combined with the electrostatic force. Assuming that the tip was contaminated with material from the surface, the short-range components were estimated using a KBr cluster as a model for the tip apex. The tip polarity could be determined by comparing the forces obtained at the maximum and at the saddle points

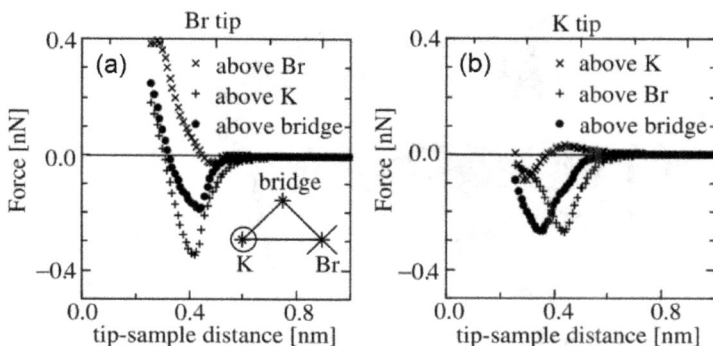

Fig. 11.2 Short-range force vs. distance curves calculated for (a) a Br^- and (b) a K^+-terminated tip above a K^+ ion, above a Br^- ion, and above a bridge position on a KBr(001) surface. Reprinted with permission from [Hoffmann *et al.* (2004)]. Copyright (2004) by the American Physical Society.

located between two neighboring cations and two neighboring anions. In the case of a K^+ terminated tip, the two force curves crossed each other, which was not the case when the tip was terminated by a Br^- ion (Fig. 11.2). On both KBr and NaCl, the calculated tip relaxation at the saddle point was much smaller than at the maximum or minimum in the images, since the tip apex mainly experiences forces generated by two positively charged ions and two negatively charged ions at about the same distance. Therefore sample relaxation becomes dominant at such locations. The shape of the force-distance curve is affected by the stronger relaxation and polarization of the anions, due to their larger size and much greater polarizability. Thus, crossing is expected for any tip which creates an electrostatic field of positive polarity around its apex.

Force-distance curves extracted from measurements on NaCl(001) were also compared to theoretical curves assuming either a MgO or a NaCl cluster as models for the tip apex [Lantz *et al.* (2006)]. Although the polarity of the tip could not be unambiguously identified from the experiments, the calculations showed that the MgO cluster induces tip jumps, in contrast to calculations for the NaCl cluster and in contrast to the measurements. Furthermore, the calculated short range attractive forces are much larger for the oxide tip than in the experiment. Thus, a stable NaCl cluster is a better model for the actual tip. Again, as the AFM tip had probably come into contact with the sample surface before the data were recorded, it is reasonable that a cluster of NaCl was adsorbed

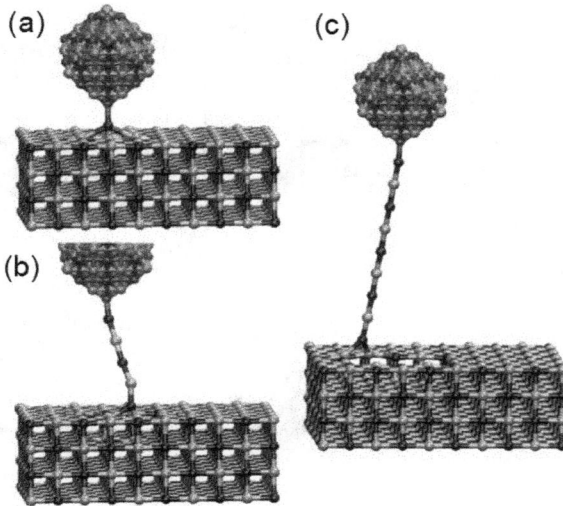

Fig. 11.3   Snapshots from simulations with an oxygen-terminated MgO tip starting above a Na^+ ion and showing several stages of NaCl chain formation at nominal tip-surface distances of (a) 0.44 nm, (b) 1.30 nm, and (c) 2.73 nm. Reprinted with permission from [Lantz *et al.* (2006)]. Copyright (2006) by the American Physical Society.

at the tip apex. No repulsion was observed in the experimental site-specific forces curves, in contrast to the computations, suggesting that the real NaCl cluster was smaller and not so symmetric. Further simulations showing the formation of a chain of alternating ions being pulled out from a NaCl(001) surface by a model oxide tip (Fig. 11.3) suggested a possible way of forming a cluster of sample material at the apex of an oxidized silicon tip.

An extension of the previous analysis consisted in mapping the potential energy landscape experienced by the AFM tip. This was obtained by integrating force-distance curves acquired on several locations on the surface under investigation [Schirmeisen *et al.* (2006)]. On NaCl(001), it was found that the tip-surface potential has a sinusoidal form with a corrugation of about 50 meV (Fig. 11.4). At specific lattice sites, discontinuities were also observed, revealing the onset of mechanical relaxation processes.

The damping signal acquired on KBr(001) was also measured as a function of the tip-surface distance [Hoffmann *et al.* (2007)]. Two regimes of distances were again identified. Above a critical separation, depending on the site, reproducible data with low noise and no interaction-induced energy

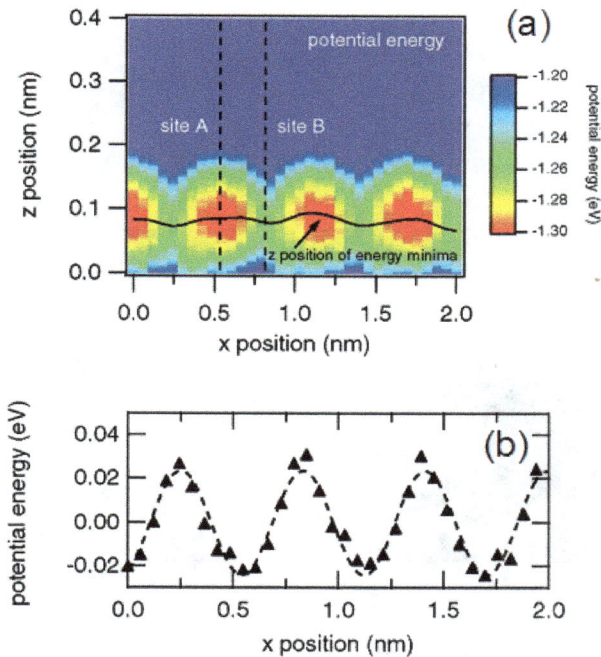

Fig. 11.4 (a) Tip-sample potential energy as a function of horizontal and vertical tip position. The black line indicates the z positions of the potential energy minima. (b) The triangles denote the horizontal line profile along the energy minima [i.e., along the black line in (a)], which can be approximated by a sinusoidal curve (dashed line), yielding an effective potential energy barrier of 48 meV. Reprinted with permission from [Schirmeisen *et al.* (2006)]. Copyright (2006) by the American Physical Society.

dissipation were measured. In this regime reproducible AFM images could be recorded. At closer tip-sample distances the frequency shift could 'jump' between two limiting curves on a timescale of tens of milliseconds. In such a case, additional energy dissipation occurred, which can be attributed to rare changes in the tip apex configuration due to short-range interactions with the sample.

11.1.2 *Alkaline earth halide surfaces*

Site specific force-distance curves on the $CaF_2(111)$ surface were also reported [Hoffmann *et al.* (2005)]. Here, high and low fluorine ions need be

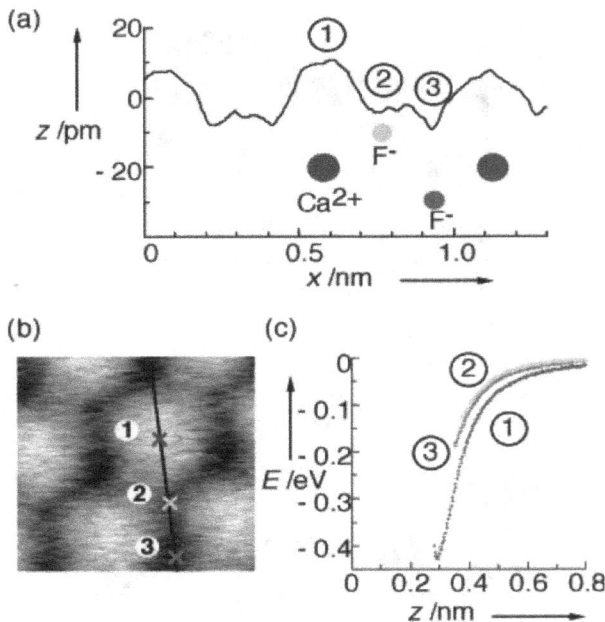

Fig. 11.5 Surface topography and site-specific force curves on $CaF_2(111)$. Selected locations above the calcium ion (1), high fluorine ion (2), and low fluorine ion (3) are shown (a) in the cross-section along the [211] direction (x) and (b) in a high-resolution image. (c) Binding energy as a function of tip-sample distance of the tip cluster at locations 1, 2, and 3 derived from experimental force curves (not shown). Reprinted with permission from [Hoffmann *et al.* (2005)]. Copyright 2005 American Chemical Society.

distinguished (Fig. 11.5). The experimental data were compared to theoretical models, in which a hydrogen-saturated Si tip with a single oxygen atom at the end, and a SiO_2 tip with hydrogen-saturated dangling bonds were considered. In the former case, a strongly repulsive interaction between oxygen and high fluorine ions was observed, resulting in an inward relaxation of the ion. In the latter case, a strong attractive interaction was found between the foremost hydrogen atom and the nearest high fluorine ions. Also on $CaF_2(111)$, by combining experimental and theoretical results it was possible to quantify the strength and distance dependence of the interaction of a tip-terminating cluster with particular surface ions and to reveal details of cluster and surface relaxation.

11.1.3 *Oxide surfaces*

To our knowledge, no experimental force-distance curves at well-defined surface locations have been reported on bulk oxide surfaces. On thin insulating films, NC-AFM spectroscopy was performed in the case of MgO grown on Ag(001) [Heyde *et al.* (2006)]. In such a case, force-distance curves with slightly different shapes could be distinguished on Mg and O sites identified by atomically resolved images.

11.2 Tunneling Spectroscopy on Thin Insulating Films

Scanning tunneling spectroscopy (STS) can be used to determine the electronic properties of ultrathin insulating films as a function of the film thickness. In the case of MgO grown on Ag(001), it was found that a thickness of only 3 monolayers already leads to the same band gap of the bulk material [Schintke *et al.* (2001)]. The surface of the trilayer has also the same electronic structure of the surface of a bulk-truncated MgO crystal, so that the electronic structure of the metal substrate is completely screened by the oxide. Tunneling spectra on CaF_2 and CaF on Si(111) surfaces could be directly compared using the stripy pattern introduced in Sec. 5.2 [Viernow *et al.* (1999a)]. The onsets of the conduction bands were associated to well-defined peaks in the normalized (dI/dV) curves. Another study on $CaF_2/Si(111)$ argued that the surface states related to the (7×7) reconstruction of the substrate are completely removed during the formation of the interface [Klust *et al.* (2005)]. However, although the band gap of CaF_2 films only 2 ML thick was essentially bulk-like, a film thickness of about 3 nm was required to fully develop the valence band structure of bulk CaF_2. An interface state was also recognized in the CaF_2 band gap close to the Fermi level. These interface states are responsible for the Moiré patterns revealed by STM on insulating films (Sec. 5.1).

The role of surface defects in STS measurements has been also addressed. Experiments on Al_2O_3 and NaCl films showed that adsorbate and defect-induced states exhibit relatively broad linewidths [Nilius *et al.* (2003); Repp *et al.* (2005b)]. Further investigations on single chlorine vacancies in ultrathin NaCl films grown on various copper substrates suggested that the broadening is caused by strong coupling between localized states and optical phonons in the film [Repp *et al.* (2005a)]. This conclusion was achieved by comparing experimental results to an inelastic resonance tunneling model with parameters taken from DFT calculations.

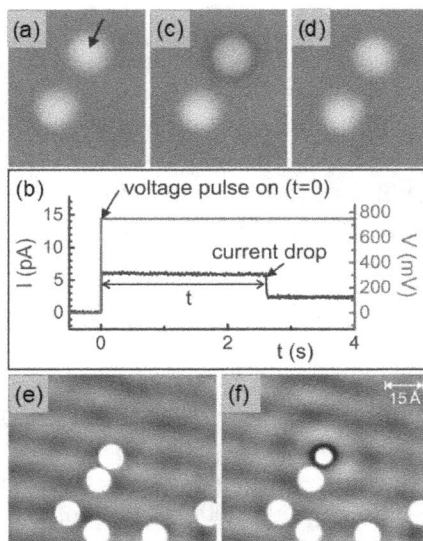

Fig. 11.6   Manipulation of a gold adatom state, as described in the text. From [Repp
et al. (2004a)]. Reprinted with permission from AAAS.

11.3   Tunneling Spectroscopy on Metal Clusters

11.3.1   *Alkali halide films*

Individual metal atoms on ultrathin alkali halide films supported by metal
surfaces exhibit different charge states, which are stabilized by the large
ionic polarizability of the film [Repp *et al.* (2004a)]. The charge state
and associated physical and chemical properties such as diffusion can be
controlled by adding (or removing) a single electron to (or from) the adatom
with a STM tip. These observations are exemplified in Fig. 11.6 for gold
atoms on a NaCl/Cu(111) system. Here, a positive pulse was applied to
the Au adatom identified by the arrow in (a). After a certain time, the
tunneling current suddenly dropped down (b), and a subsequent image (c)
revealed a sombrero-like shape of the atom. The original state of the atom
was recovered by applying a negative pulse (d). The two different states of
the Au adatoms were further characterized using the 2D electron gas in the
interface state band of NaCl/Cu(111). The adatoms in the initial state (e)
did not scatter interface-state electrons, whereas the manipulated adatom
(f) acted as a scatterer.

11.3.2 *Oxide films*

Different adsorption sites of metal atoms deposited on thin alumina films could be recognized from variations in the electronic states of the adatoms [Nilius *et al.* (2003)]. The adsorption of gold atoms is, once again, particularly interesting. In such a case, self-assembled gold chains with a maximum length of 2.25 nm, and a preferential orientation close to the [001] direction of the underlying NiAl(110) substrate, were observed [Kulawik *et al.* (2006)]. Fig. 11.7 summarizes STS data acquired on a gold pentamer. By increasing the absolute value of the sample bias, the pentamer or parts of it were desorbed. At negative biases, two resonances were observed [Fig. 11.7(a)]. These resonances had the same symmetry and were located at each of the five lobes and almost vanished inbetween. At positive biases, three states were identified, each of them localized at specific parts of the chain [Fig. 11.7(b)].

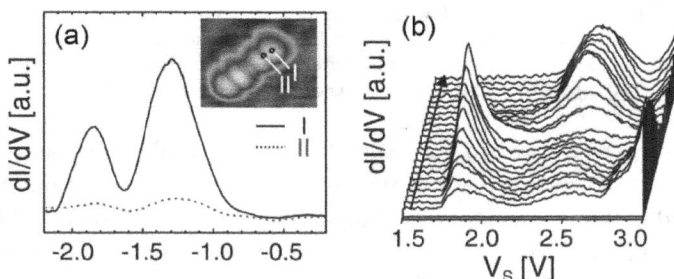

Fig. 11.7 Conductance spectra of a gold pentamer on an alumina film formed on NiAl(110). (a) At negative sample bias V_s, two states are detected, which are localized at each of the five lobes visible in the inset (position I). In between the lobes, nodal planes with strongly reduced conductance are observed (position II). (b) Conductance spectra taken at positive V_s differ along the axis of the Au chain. Reprinted with permission from [Kulawik *et al.* (2006)]. Copyright (2006) by the American Physical Society.

11.4 Tunneling Spectroscopy on Organic Molecules

An investigation on phthalocyanine derivatives adsorbed on an ultrathin alumina film grown on a NiAl(110) surface revealed a series of equally spaced features in the differential conductance spectra, which were attributed to vibrational states of the individual molecules [Qiu *et al.* (2004)]. The coupling of the electron current to the vibrational modes depended on the structure of the adsorbed species. When the same

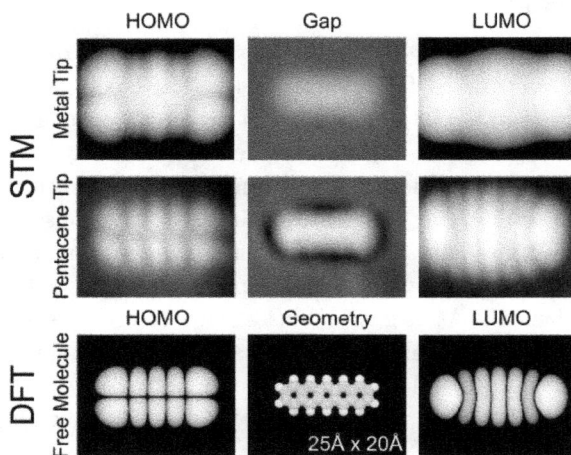

Fig. 11.8 STM images acquired with a metal and a pentacene tip, and contours of constant orbital probability distribution of a free pentacene molecule. The geometry of the free molecule is displayed in the lower center image. Reprinted with permission from [Repp *et al.* (2005b)]. Copyright (2005) by the American Physical Society.

molecules were adsorbed on the bare NiAl(110) substrate, surface vibrational features were not observed. Vibrational spectroscopy was also performed on C_{60} molecules disentangled from a metal substrate by means of ultrathin NaCl films. A series of images showing the electronic structure of a pentacene molecule on NaCl/Cu(111) is shown in Fig. 11.8 [Repp *et al.* (2005b)]. The spatial resolution significantly improved when a molecule was picked up by the STM tip. The STM images for bias voltages in the HOMO-LUMO band gap (center) appeared relatively featureless, whereas the images at bias voltages exceeding the HOMO (left) or LUMO (right) exhibited very pronounced features, resembling the electron density of the HOMO and LUMO of the free molecule. The probing tip can also be used to induce transitions in the state of organic molecules. This has been shown in recent measurements on naphthalocyanine, where sudden 'switching' in the molecular conductance could be induced by STM at low temperature [Liljeroth *et al.* (2007)].

Chapter 12

Nanotribology on Insulating Surfaces

Understanding and controlling friction on the nanoscale is one of nowadays' challenges for scientists and engineers. Since the first observation of atomic friction, reported for a tungsten tip sliding on graphite [Mate *et al.* (1987)], significant progress has been made in this topic. An accurate description of the motion of a nanotip driven on a crystal surface by a microcantilever is given by the *Tomlinson model* [Tomlinson (1929); Tomanek *et al.* (1991)]. According to this model, the tip sticks to a given equilibrium position on the surface until the driving force becomes high enough to induce a jump into the next equilibrium position. The consequent *stick-slip* motion reflects the atomic periodicity of the surface with a saw-tooth like modulation of friction. However, this scenario is observed only if a precise condition is verified. The elastic constant of the cantilever spring must be lower than the curvature of the tip-surface potential. If this condition is not satisfied, the tip will move on the surface in a continuous way, entering a so-called *superlubric* regime. Alkali halide surfaces have been an important benchmark to test the Tomlinson model and its further extensions. Atomic stick-slip and the transition to the superlubric regime can be readily observed on these surfaces in UHV. A logarithmic velocity dependence of friction was also recognized on these surfaces, and attributed to thermal activation of the tip jumps. On alkali halide surfaces abrasion wear and nanoindentation can be also investigated down to the atomic scale, opening interesting scenarios in nanotribology.

12.1 Friction Mechanisms at the Atomic Scale

12.1.1 *The Tomlinson model*

Let us consider a nanotip pulled across a periodic lateral potential by a spring attached to a support moving with a constant speed v (Fig. 12.1). The spring stiffness, $k_{L,\text{eff}}$, combines the lateral stiffnesses of driving support and contact region (Sec. 3.1.2). For sake of simplicity, we assume that the potential describing the tip-surface interaction has a sinusoidal shape with amplitude U_0 and periodicity a. We also distinguish between the tip position, x, and the position of the moving support, $x_s = vt$. The latter quantity, and not the tip position, is recorded in FFM experiments. Provided that the scan speed v is sufficiently low, the tip resides in a minimum of the total potential

$$U(x,t) = U_0 \cos\left(\frac{2\pi x}{a}\right) + \frac{1}{2}k_{L,\text{eff}}(x - vt)^2.$$

As shown in the next paragraph, the movement of the tip from one minimum to the next one can be continuous or not, depending on the relation between the tip-surface potential and the elastic energy stored in the spring. This relation can be quantified by the parameter

$$\eta = \frac{4\pi^2 U_0}{k_{L,\text{eff}}a^2}. \tag{12.1}$$

If $\eta < 1$ the tip motion is continuous. If $\eta > 1$, stick-slip occurs. By analyzing the conditions of equilibrium for the tip position, one also finds that the amplitude of the tip-surface potential U_0 is proportional to the maximum lateral (*friction*) force $F_{L,\text{max}}$, according to the relation [Socoliuc *et al.* (2004)]

$$U_0 = \frac{aF_{L,\text{max}}}{2\pi}. \tag{12.2}$$

The effective spring constant $k_{L,\text{eff}}$ is related to the slope k_{exp} of the lateral force curve as a function of the displacement x_s (at the onset of the support motion) by the relation

$$k = \left(1 + \frac{1}{\eta}\right)k_{\text{exp}}. \tag{12.3}$$

Combined with the definition (12.1), the relations (12.2) and (12.3) allow to extract the parameter η from frictional force maps recorded in FFM measurements.

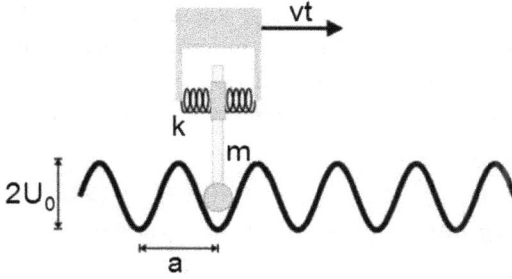

Fig. 12.1 Sketch of the Tomlinson mechanism. A point mass m, representing a FFM tip, is driven by an elastic spring across a sinusoidal energy landscape.

12.1.2 *Superlubricity*

Simple analytical considerations lead to the following relations for the critical position at which the tip equilibrium becomes unstable, and for the corresponding value of the friction force [Gnecco *et al.* (2001)]:

$$x_c = \frac{a}{2\pi} \arccos\left(-\frac{1}{\eta}\right), \qquad F_c = \frac{k_{L,\text{eff}} a}{2\pi} \sqrt{\eta^2 - 1}. \qquad (12.4)$$

When the critical position is reached, the tip suddenly 'jumps' into the next equilibrium position, and a finite amount of energy ΔU is released into the underlying surface. Eq. (12.4) clearly shows that the stick-slip can only occur if $\eta > 1$. If $\eta < 1$, i.e. if the contact is too stiff or the tip-sample interaction is too weak, the tip equilibrium is always stable, and energy cannot be dissipated via sudden tip jumps. In what follows, we will refer to this case as the *superlubric regime*. In the limit case $\eta \to 1$, the energy dissipation ΔU is related to the friction parameter η by $\Delta U \propto (\eta - 1)^2$ [Baratoff *et al.* (2009)]. Thus, the transition from stick-slip to superlubricity is smooth, i.e. the dissipative mechanism vanishes when $\eta \to 1$ without abrupt variations. In this sense, we can assume that the friction parameter η behaves like the order parameter in a second-order phase transition. In the opposite limit $\eta \to \infty$, the linear relation $\Delta U \propto \eta$ holds.

The normal force corresponding to the superlubricity onset is usually in the order of the instrumental noise, which makes the superlubric regime quite difficult to control. However, a state of ultralow friction can also be achieved if the normal force oscillates around higher values, provided that the oscillation amplitude is high enough [Socoliuc *et al.* (2006)]. As a first approximation, we can assume that the amplitude U_0 of the tip-surface potential is replaced by $U_0(1+\alpha\cos\omega t)$, where ω is the oscillation frequency. If $\omega \gg v/a$, the tip experiences the minimum corrugation several times as the cantilever slowly crosses the distance separating adjacent potential minima, and it slides without jumping if the less restrictive condition $\eta(1-\alpha) < 1$ is satisfied.

12.1.3 *Velocity dependence of atomic friction*

The motion of a nanotip is also influenced by thermal effects. As a result of thermal vibrations, the tip may jump when the energy barrier between two adjacent potential minima is in the order of $k_B T$. Using reaction rate theory, it can be proven that, in the stick-slip regime, the following relation between the friction force F_L, the temperature T and the scan velocity v approximately holds:

$$F_L(T, v) = F_{L,\text{max}} + \frac{k_B T}{\lambda}\ln v, \qquad (12.5)$$

where v is expressed in nm/s and λ is a constant in the order of 1 nm [Gnecco *et al.* (2000)]. Thus, at a fixed temperature, the friction force increases logarithmically with the sliding velocity. Relation (12.5) is not valid if the tip jump occurs very close to the critical point $x = x_c$, which is the case if the scan speed is sufficiently high. In such a case a friction 'plateau' is reached, corresponding to $F_L = F_{L,\text{max}}$ [Riedo *et al.* (2003)]. In the opposite case of $v \to 0$, back and forward jumps occur, leading to vanishing of the average value of the friction force (*thermolubricity*) [Krylov *et al.* (2005)].

12.2 Friction on Halide Surfaces

12.2.1 *Friction on bulk halide surfaces*

Friction on alkali halide surfaces has been systematically investigated using FFM setups in UHV [Gnecco and Meyer (2007)]. Friction maps on the atomic scale usually reveal periodicities corresponding to the distance of equally charged ions (Fig. 12.2). The corresponding friction profiles exhibit the stick-slip phenomenon discussed in the previous section.

The KBr(001) surface tends to exhibit very low friction in UHV (friction coefficient < 0.04) [Lüthi *et al.* (1996)]. However, wear of this surface is typically observed when the load exceeded 3 nN, resulting in a sudden

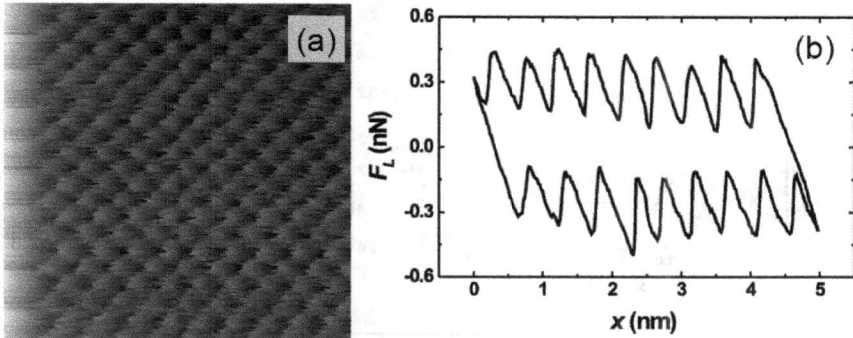

Fig. 12.2 (a) Lateral force map (5 nm) on a NaCl(001) surface and (b) friction loop formed by two scan lines acquired while scanning forwards and backwards. The normal force is $F_N = 0.65$ nN. Reprinted from [Gnecco *et al.* (2000)]. Copyright (2000) by the American Physical Society.

increase of the friction coefficient up to 1.2. Wear on KBr surfaces has been well documented by high resolution images, as shown in the next section. The transition from stick-slip to superlubricity predicted by the Tomlinson model was first observed on NaCl(001) [Socoliuc *et al.* (2004)]. When the normal load was reduced below 1 nN, the hysteresis loop and the accompanying dissipation disappeared within the sensitivity of the experimental setup (Fig. 12.3). At the same time, the saw-tooth modulation of the lateral force was transformed into a continuous modulation of perfect match between forward and backward scan, still showing the atomic periodicity of the surface lattice. In another experiment, friction could be 'switched' on and off by exciting mechanical resonances of the contact volume perpendicular to the sliding direction [Socoliuc *et al.* (2006)]. The corresponding variation of the interaction energy reduced friction below the instrumental noise level without any noticeable wear, as discussed in the previous section. Without actuation, atomic stick-slip and dissipation were observed. The state of *dynamic superlubricity* so achieved has been also exploited to extend the resolution capabilities of contact AFM [Gnecco *et al.* (2009)].

In another series of measurements atomic stick-slip on KBr(001) was recorded with a high bandwidth of 3 MHz, showing a wide variation of slip durations up to several milliseconds [Maier *et al.* (2005)]. These slip events are by far longer than expected for typical relaxation processes on atomic scale, and were attributed to multiple contacts between tip

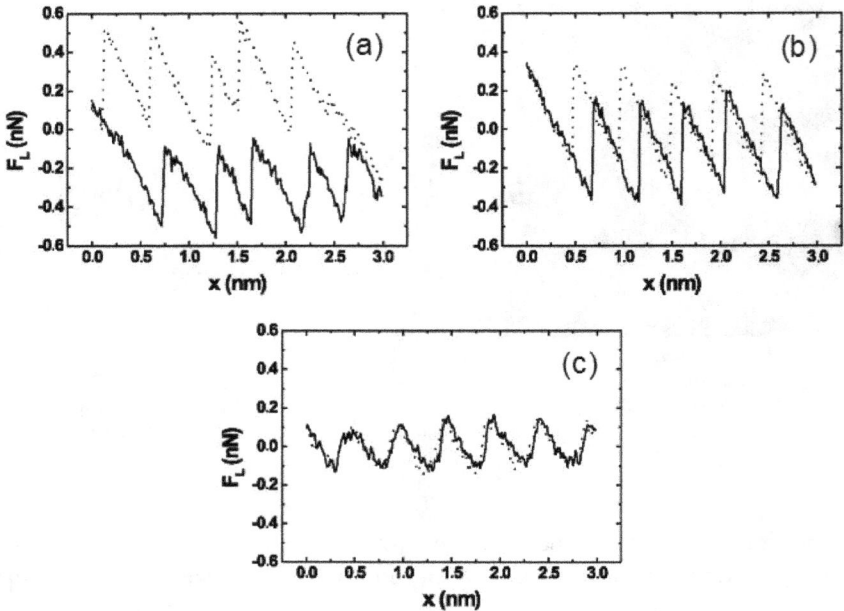

Fig. 12.3   (a)-(c) Friction loops acquired on a NaCl(001) surface with normal forces of
(a) 5.4 nN, (b) 4.0 nN, and (c) 0.2 nN. The hysteresis loop is reduced till it vanishes in the
superlubric regime. Reprinted with permission from [Socoliuc *et al.* (2004)]. Copyright
(2004) by the American Physical Society.

and surface.   This conclusion was supported by a comparison of the
experimental results with a multi-tip simulation based on the Tomlinson
model at finite temperature. Thermal effects were also recognized in friction
measurements on NaCl(001), compatible with the logarithmic velocity
dependence expected from Eq. (12.5) [Gnecco *et al.* (2000)].

We should also mention that the energy landscape sensed on atomically
flat surfaces is modified at surface steps, where friction increases (Fig. 12.4).
This result is due to increased energy barriers at these locations, otherwise
known as *Schwöbel barriers* [Meyer *et al.* (1996); Hölscher *et al.* (2008)].

12.2.2   *Friction on halide films*

Friction force microscopy has been also applied to thin insulating films.
The 'rumpling' effect induced by the growth of KBr films on NaCl (Sec. 5.3)
appeared quite clearly in highly resolved FFM images [Maier *et al.* (2008b)].

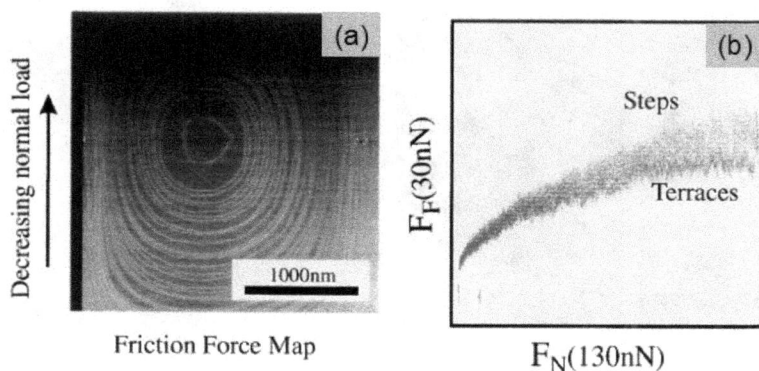

Fig. 12.4 (a) Friction force map on NaCl(001). The loading is decreased from 140 to 0 nN (jump off point) during imaging. (b) Two dimensional histogram of (a). The data pile up in distinct regions, corresponding to terraces and step edges. Reprinted with permission from [Meyer *et al.* (1996)]. Copyright 1996, American Institute of Physics.

As shown in Fig. 12.5, a spatial modulation of friction could be observed only when $\eta > 1$, while rumpling disappeared at the superlubricity onset. Considering that the corrugation of the ionic monolayers is highest at the interface between KBr and NaCl (Fig. 5.12), this result strongly suggests that contact AFM is sensitive to subsurface inhomogeneities. A slight modulation of the friction force was also reported on superstructures formed by ultrathin KBr films grown on a Cu(001) surface [Filleter *et al.* (2008)]. Material contrast (without lattice resolution) was reported in FFM images of CaF_2 overgrowing a CaF layer on a Si(111) surface [Klust *et al.* (1998)]. In such case, friction was found to be higher on CaF than on CaF_2.

12.3 Nanowear Processes on Insulating Surfaces

12.3.1 *Abrasion wear on alkali halide surfaces*

Friction force microscopy can be also used to study wear processes on the nanoscale. Once again, the simple structure of alkali halide surfaces makes them very suitable for this kind of studies. In a model experiment, a surface of KBr(001) was scratched by bringing the tip into contact, then scanning forth and back, and increasing the normal load until some atomic layers were removed from the substrate [Gnecco *et al.* (2002)]. The time evolution of the lateral force under scratching reveals that the wear mechanism on

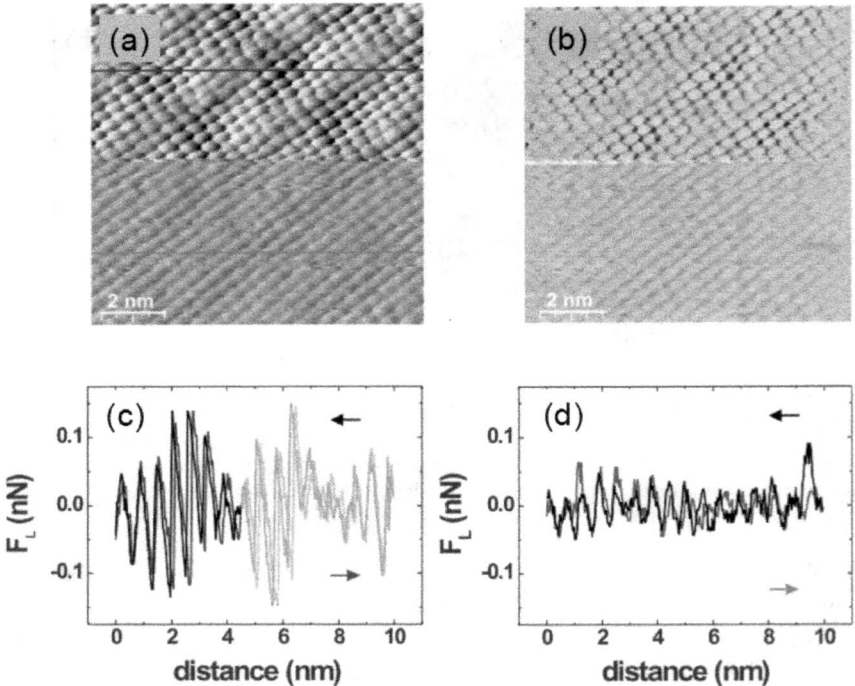

Fig. 12.5 (a) Lateral force maps (10 nm) of the superstructure formed by a thin KBr film on NaCl(001). The average normal force was reduced from 0.36 nN to 0.05 nN in the lower half of the image. (b) Difference between lateral forces recorded in successive forward and backward scans. (c, d) Forward and backward traces corresponding to the lines indicated in (a). Reprinted with permission from [Maier *et al.* (2008b)]. Copyright (2008) by the American Physical Society.

alkali halides is due to removal and rearrangement of single ion pairs. The material at the end of the scratch, i.e. in front of the moving tip, perfectly recrystallized in atomically flat terraces, which exhibited exact atomic registry with the underlying substrate lattice, as in local epitaxial growth (Fig. 12.6). The applied load had a strong influence on the abrasion process, whereas the scan velocity did not play any major role.

It was also observed that, once an atomic layer was damaged at some specific spot on KBr(001), the whole layer was removed and the debris reorganized in a process accompanied by strong, irregular friction. After some tens of scan lines a relatively ordered ripple structure was formed across the scan lines, which is shown in Fig. 12.7(a) [Socoliuc *et al.*

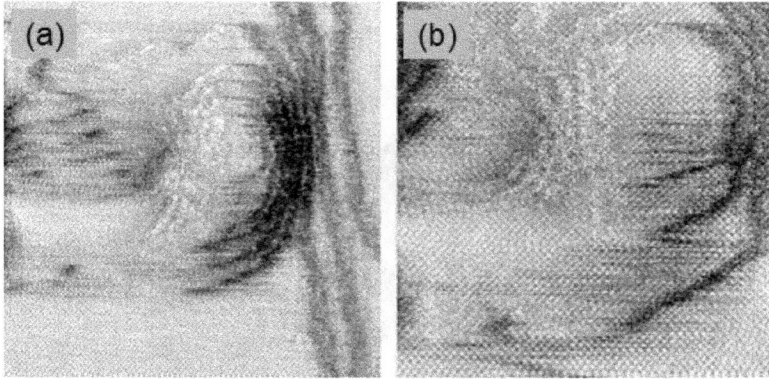

Fig. 12.6   (a) Lateral force map (39 nm) acquired at the end of a groove created after scanning a straight line for 256 times with $F_N = 21$ nN. (b) Zoom-in (25 nm) showing the atomic structure of the mound resulting from the scratching process. Reprinted from [Gnecco *et al.* (2002)]. Copyright (2002) by the American Physical Society.

(2003)]. Some understanding of the process can be obtained by comparing the corrugated topography and the lateral force profiles simultaneously acquired in the wear process [Fig. 12.7(b)]. The ripples emerged when material transported in front of the tip increased the friction, slowing down the tip movement until the lateral force was strong enough to make the tip jump over the mound of material. During following scan lines, existing mounds collect further material and grow up. In this scenario the typical distance between two ripples should be in the order of the tip size, which is in agreement with experimental results obtained with different tips. When scanning across a square area, the ripple pattern coalesce giving rise to structures like those shown in Fig. 12.8.

12.3.2   *Nanoindentation processes*

Nanoindentation processes have been also characterized on alkali halide surfaces in UHV. The same AFM tips could be used for both damaging and imaging the surface in non-contact mode. Fig. 12.9(a) shows monatomic terraces resulting from dislocation nucleation and glide on a KBr(001) surface [Filleter *et al.* (2006)]. The discontinuities in the force-distance curves recorded during indentation [Fig. 12.9(b)] correspond to the creation of dislocation loops in the crystal. The nucleation of a dislocation caused an abrupt monatomic layer displacement of the tip into the sample and

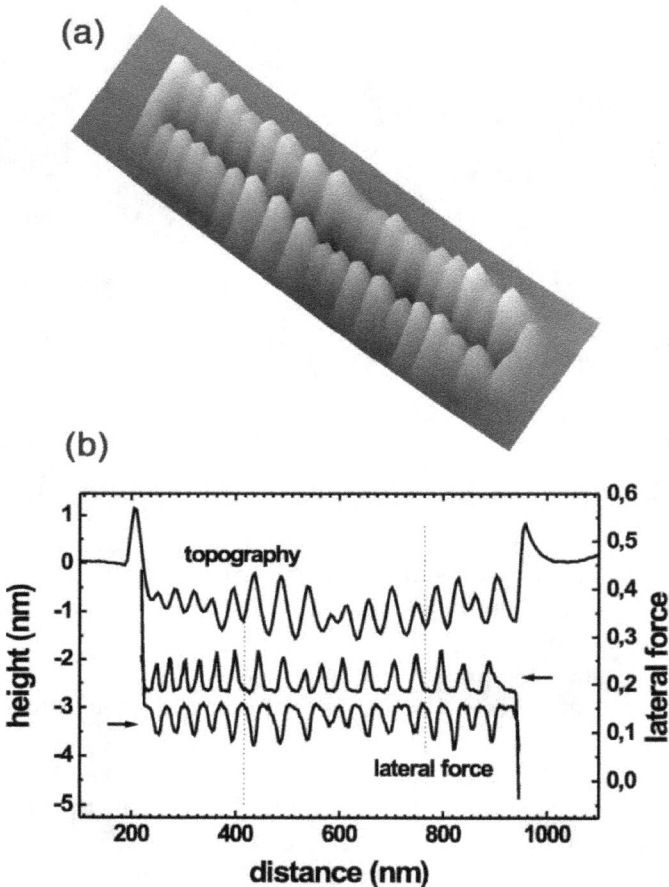

Fig. 12.7   (a) Topography image of a groove formed on KBr(001) after 512 scans along the (100) direction of the crystal surface (with $F_N = 27$ nN). (b) Cross section of the topography along the ground of the groove and lateral force acquired while scratching the groove. The dotted lines indicate positions in the pits where the tip is decelerated as derived from an increase of the lateral force in either direction. Reprinted from [Socoliuc *et al.* (2003)]. Copyright (2003) by the American Physical Society.

the consequent creation of monatomic terraces. The lateral extent of the dislocation structure and the distribution of force discontinuities during indentation were significantly influenced by the tip radius. The shear stress at the yield point, as determined from the experiment using continuum mechanics, turned out to be in the order of 2.5 GPa, which is in agreement

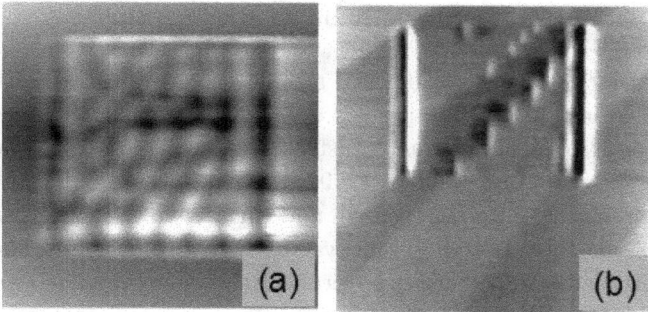

Fig. 12.8 Contact mode AFM images of square regions (500 nm large) on KBr(001) surfaces previously scanned with high load. (a) Originally flat region. (b) Region with cleavage steps running from the lower left to the upper right corner of the frame. Reprinted from [Socoliuc *et al.* (2003)]. Copyright (2003) by the American Physical Society.

with theoretical predictions for the ideal shear stress of KBr [Ogata *et al.* (2004)]. Nanoindentations were also performed on a MgO(001) surface by a combination of contact and tapping mode AFM [Arce *et al.* (2000)]. In such a case, discrete discontinuity events were also observed in the applied force-distance curve during indentation, but no dislocation structure could be recognized in the surface topography.

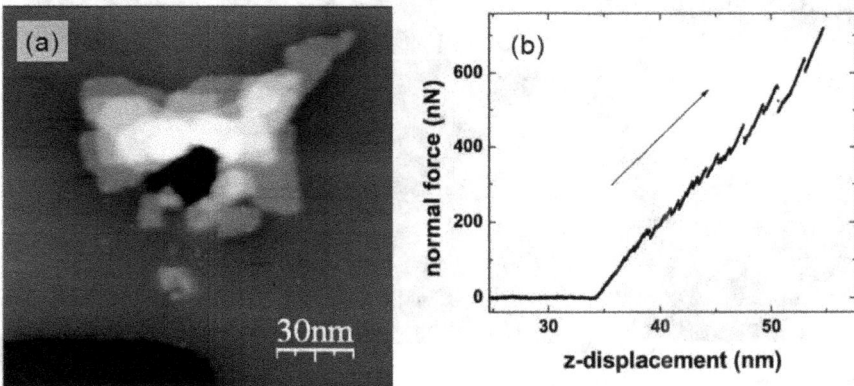

Fig. 12.9 (a) NC-AFM image of a KBr(001) surface after indentation with a sharp tip (7 nm radius). Four layers of monatomic terraces are piled up around the indentation pit, while the lateral extension is still limited to a radius of 50 nm. (b) Force-distance curve recorded while indenting the surface. Reprinted with permission from [Filleter *et al.* (2006)]. Copyright (2006) by the American Physical Society.

12.4 Modeling Nanotribology on Insulating Surfaces

An early theoretical investigation of atomic friction, dealing with a MgO tip sliding over a LiF(001) surface, revealed a regular stick-slip behavior of the lateral force only after transfer of several ions from the sample to the tip [Livshits and Shluger (1997)]. This ion transfer established a higher commensurability between tip and surface, resulting in a significant decrease of the total energy of the system. More recently, atomic friction on KBr(001) was studied using several KBr clusters simulating the apex of an AFM tip [Wyder *et al.* (2007)]. At each time step the tip was displaced over small distances, and for each displacement the ions were fully relaxed to their respective equilibrium positions and the normal and lateral forces were subsequently determined. The good agreement with FFM measurements suggests that atomic friction experiments should involve

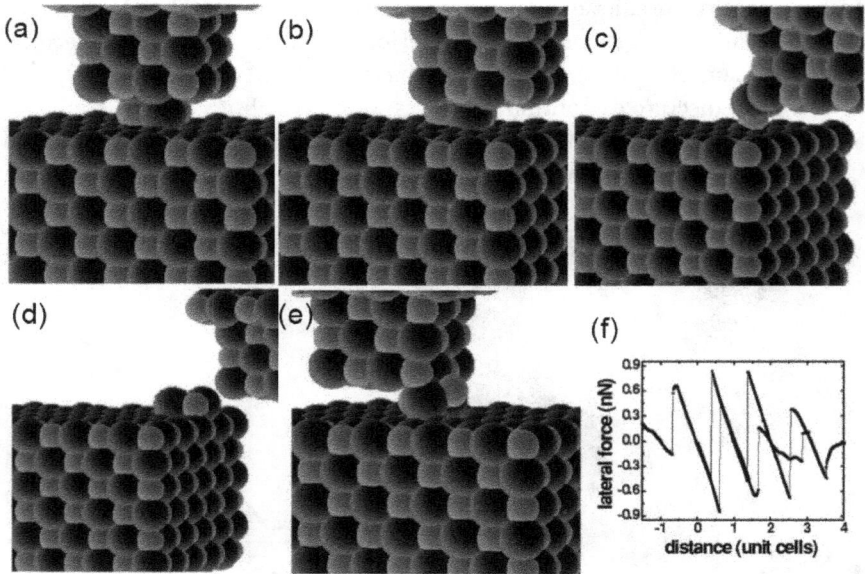

Fig. 12.10 (a-e) Atomistic simulation of sliding on KBr(001) using a 'cube on flat' tip model. (a) Starting configuration. (b-c) Rotational jumps of the lowermost tip layer upon lateral movement. (d) The lowermost tip layer is left behind the tip on the surface. (e) The returning tip picks up the island and then pushes it under rotational jumps along the sliding direction. (f) Lateral force data from the simulation exhibiting characteristic stick-slip features. Reprinted with permission from [Wyder *et al.* (2007)]. Copyright 2007, American Institute of Physics.

contact sizes of only a few atoms. Furthermore, the occurrence of stick-slip instabilities seems to depend on the compliance of the tip rather than on properties of the substrate, suggesting that on hard crystalline surfaces, FFM is primarily a method to study dissipation processes in the sliding tip. Imperfect commensurability between the tip apex and the surface produces irregularities in the lateral force curves on length scales that are smaller than the unit cell of the substrate (Fig. 12.10).

The conspicuous number of atoms and the time-scales which characterize nanowear processes make computer simulations of these phenomena quite challenging. Among them, we would like to mention an interesting phenomenon, which has been predicted on NaCl(001) close to the substrate melting point, T_M [Zykova-Timan *et al.* (2007)]. Two distinct and opposite effects were observed for ploughing and for grazing friction in this regime. A frictional drop close to T_M was found for deep ploughing and wear, but on the contrary a frictional rise was observed for grazing, wearless sliding. The former effect could be related to 'skating' through a local liquid cloud (Fig. 12.11), whereas the latter effect was due to linear response properties of the free substrate surface.

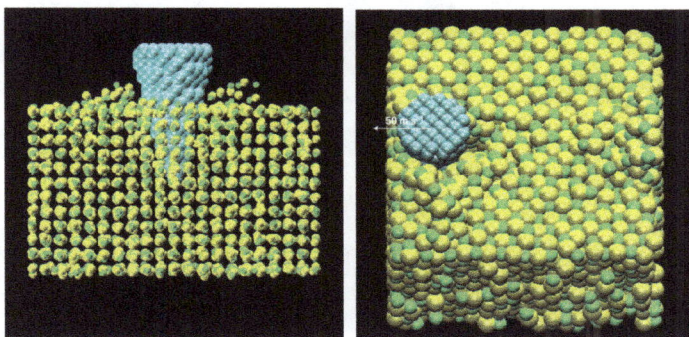

Fig. 12.11 At the melting point, a hard tip ploughing through solid NaCl(001) is surrounded by a local liquid cloud that moves with it. The light blue, green and yellow spheres represent tip atoms, Na^+ and Cl^-, respectively. Note the fast-healing furrow behind the tip. The penetration depth is 1.8 nm. Reprinted by permission from Macmillan Publishers Ltd: Nature Materials [Zykova-Timan *et al.* (2007)], copyright 2007.

Chapter 13

Nanomanipulation on Insulating Surfaces

Detecting and controlling nanomanipulation processes is essential in the development of molecular electronics and quantum computing, as well as to investigate elementary surface processes. The STM technique is well-established in this context, but its application is limited to conducting systems. On the other side, contact AFM can be used for pushing and pulling nano-objects, but in order to control the manipulation of surface atoms with the precision achieved by STM on metals, AFM has to be used in non-contact mode. Although promising results came from semiconducting surfaces, where atomic manipulation was reported [Sugimoto *et al.* (2005)], NC-AFM manipulation on insulating surfaces remains quite challenging and, up to the time of writing, only small adsorbates and vacancies has been moved with limited degree of control and reproducibility on these substrates. However, a few attempts to simulate nanomanipulation of adatoms and vacancies by 'virtual' AFM suggest feasible strategies for future experiments.

13.1 Nanomanipulation Experiments on Insulating Surfaces

In order to perform nanomanipulation by SPM, the probing tip has to be brought to a distance from the surface that is closer than in typical imaging conditions, so that the tip apex interacts more strongly with the object to be moved. The occurrence of a manipulation event depends on the chemical nature and the atomic structure of the tip, and its evolution is influenced by the temperature and the tip trajectory. In any case, manipulation events cause sudden changes in the force field experienced by the tip, and their influence on the cantilever motion, as well as the response of the instrumentation, are quite difficult to predict. In NC-AFM, evidence of successful manipulation events may be given by sudden changes in the frequency shift and damping signals, provided that these signals remain well above the experimental noise level.

13.1.1 *Manipulation on halide surfaces*

Nanomanipulation of C_{60} islands on a NaCl(001) surface was reported in an early study by contact AFM [Lüthi *et al.* (1994)]. The C_{60} islands could be both translated and rotated by the tip, as shown in the series of snapshots in Fig. 13.1. An extremely small dissipation energy of 0.25 meV per molecule and a cohesive energy of 1.5 eV were estimated from the lateral force signal acquired simultaneously with the topography. The corresponding shear strength (less than 0.1 MPa) was much lower than typical values observed for boundary lubricants, which suggested a possible application of the C_{60} islands as 'nanosleds' for large organic molecules. It is interesting to note that a similar experiment on a graphite surface resulted in complete disruption of the C_{60} islands.

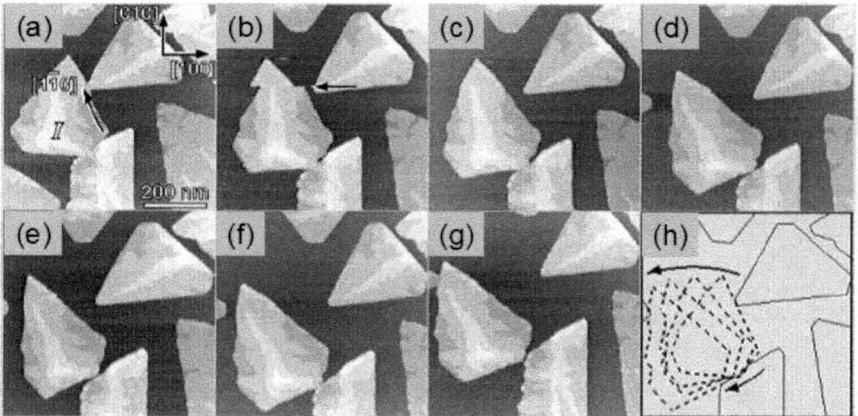

Fig. 13.1   (a-g) Sequence of AFM images (530 nm) of a C_{60} island manipulated on a NaCl(001) surface by contact AFM. (h) Overview of the island motion. From [Lüthi *et al.* (1994)]. Reprinted with permission from AAAS.

More recently, two studies focused on small defects on $CaF_2(111)$ manipulated by NC-AFM [Hirth *et al.* (2006); Fujii and Fujihira (2007)]. The defects were produced by exposing the substrate to the residual gas of

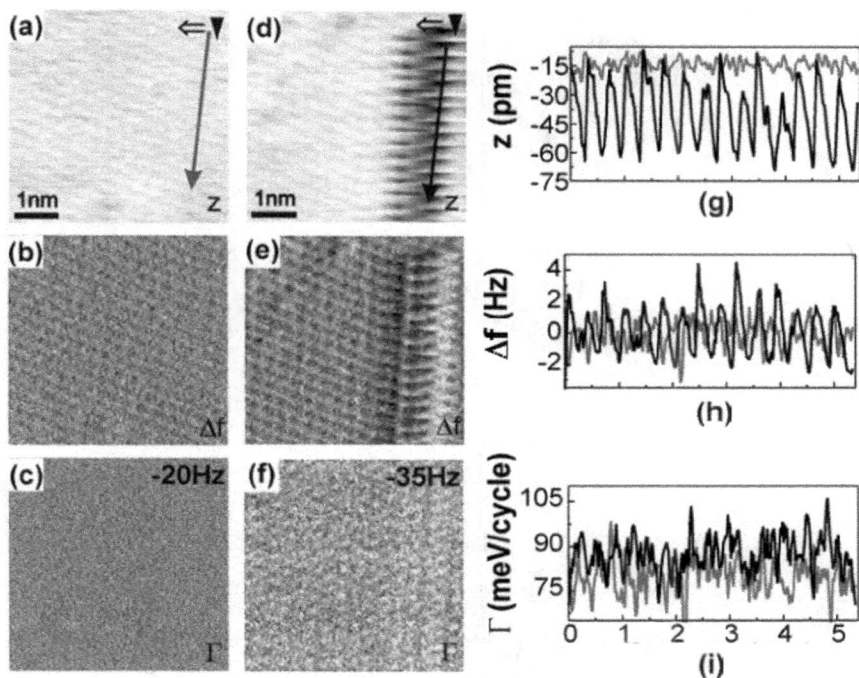

Fig. 13.2 (a, d) Topography, (b, e) frequency shift and (c, f) damping maps acquired on a CaF$_2$(111) surface (first column) in imaging mode ($\Delta f = -20$ Hz) and (second column) in manipulation mode ($\Delta f = -35$ Hz). (g-i) Profiles corresponding to the cross-sections in (a) and (d). From [Hirth *et al.* (2006)]. Reproduced with permission from IOP Publishing.

the UHV chamber (essentially water and carbon monoxide), or by gently crashing the AFM tip into the surface. In the first case the defects were presumably pushed by the tip, and a kind of stick-slip behavior was observed (Fig. 13.2). Considering the higher adsorption energy of polar H$_2$O molecules, compared to other chemical species, these defects were associated to water. In the second case, a point defect was imaged at decreasing tip-sample distances till the defect suddenly jumped into one of the six neighboring sites. Comparing the atomic contrast produced by charged tips with different terminating ions (Sec. 4.1), and the lateral mobility of the defect, the defect could be associated to a positively charged hydrogen ion in a water molecule occupying a Ca^{2+} site. Stick-slip motion was also reported when vacancies on a KCl(001) surface were laterally moved by NC-AFM [Nishi *et al.* (2006)]. From the images acquired in

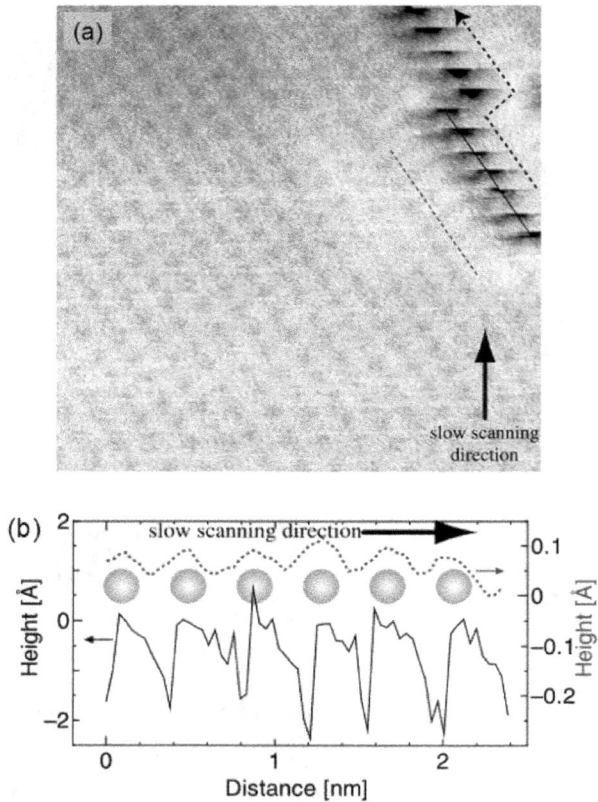

Fig. 13.3   (a) NC-AFM image (7.6 nm) of a KCl(001) surface in manipulation mode ($\Delta f = -10.5$ Hz). In the upper right part, a vacancy moves. (b) Line profile of the solid line in Fig. 13.3(a) and average profile of 12 lines equivalent to the dotted line. From [Nishi *et al.* (2006)]. Reproduced with permission from IOP Publishing.

the manipulation, it appears that the nearest neighbors of the vacancies were pulled rather than pushed by the tip into the vacancies (Fig. 13.3).

13.1.2   *Manipulation on oxide surfaces*

Manipulation of Pd clusters on MgO(001) was reported, but not further investigated, in a contact AFM study with normal loads below 10 nN [Haas *et al.* (2000)]. Palladium adatoms were manipulated by STM on Al_2O_3 [Nilius *et al.* (2003)] and on Mg(001) thin films [Sterrer *et al.* (2007)] by increasing the tunneling current or by applying a voltage pulse. The adatoms suddenly jumped into neighbor sites or completely disappeared from the areas under investigation.

13.2 Modeling Nanomanipulation on Insulating Surfaces

Computer simulations of adsorbates and vacancies manipulated by NC-AFM appeared only recently in literature. Here, we need to distinguish between 'true' manipulation events induced by the probing tip, and thermal diffusion of impurities, as imaged by NC-AFM.

13.2.1 *AFM imaging of surface diffusion*

Palladium atoms on MgO(001) surfaces were recently chosen as model systems to investigate NC-AFM imaging of adsorbates freely diffusing on insulating substrates [Watkins *et al.* (2007)]. The nanotip consisted in a MgO cube, and imaging was performed in constant-height mode. A small contrast in the damping signal was predicted when the adsorbates diffuse on the surface, even in absence of dissipation mechanisms due to tip-induced structural changes [Fig. 13.4(a)]. When the atoms become immobile (at lower temperatures), the contrast in the damping signal was expected to vanish. The diffusing impurities should also produce a characteristic noise in the frequency shift images, which is almost indistinguishable from other sources of experimental noise [Fig. 13.4(b)]. At intermediate temperatures, when the Pd atoms diffuse slowly and the time between

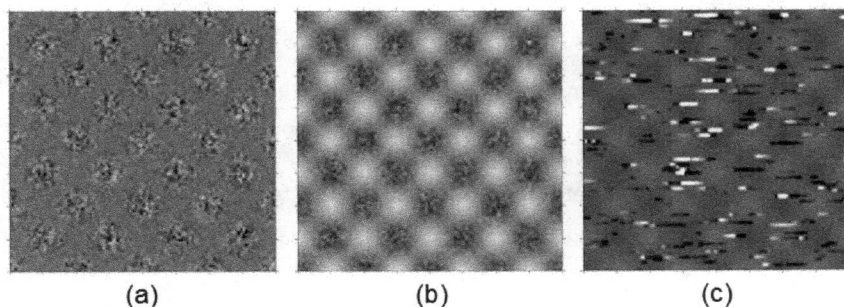

(a)   (b)   (c)

Fig. 13.4   Simulated (a) damping and (b) frequency shift images (1.7 nm) of a MgO(001) surface with a Pd adatom diffusing at 150 K. (c) Frequency shift image of the same system at 110 K. Reprinted with permission from [Watkins *et al.* (2007)]. Copyright (2007) by the American Physical Society.

jumps is comparable to the time of a line scan, stripes are expected in the Δf images [Fig. 13.4(c)]. By analyzing the length distribution of these stripes, it should be possible to determine the average escape rate at a given temperature. A similar analysis was also conducted on a water molecule rotating about an oxygen atom on a $CeO_2(111)$ surface [Watkins *et al.* (2007)].

13.2.2 *Nanomanipulation of adatoms and vacancies*

The influence of the AFM tip on the motion of Pd atoms on a Mg(001) surface has been the subject of another theoretical investigation [Trevethan *et al.* (2007b)]. Since the Pd adatoms adsorb on oxygen atoms in the substrate [Xu *et al.* (2006)] and they can jump into one of the four equivalent nearest-neighbor oxygen sites, four barrier fields had to be introduced in the Monte Carlo algorithm used to reproduce the tip motion. When a jump occurred, both the force field and the barrier fields were immediately shifted to the new position of the adatom, without interrupting the computer simulation. The jump caused hysteresis in the tip-surface interaction over a single oscillation cycle of the cantilever, which led to a spike in the corresponding damping signal.

Nanomanipulation of oxygen vacancies on MgO(001) has been also simulated [Trevethan *et al.* (2007c)]. In such case, when the tip was brought close enough to the surface, oxygen atoms were found to jump into neighboring vacancies. Fig. 13.5 shows the simulated frequency shift, closest approach, and damping signal at different temperatures as a function of the cantilever oscillation cycle number. When an oxygen ion jumps into a vacancy or even onto the tip apex, a spike in the frequency shift over the following cycles appears, due to the sudden change of the force field experienced by the tip. The jump is followed by a spike in the excitation signal which is caused by the energy lost by the cantilever, and the excitation finally decays. The whole process is strongly temperature-dependent. If the temperature increases, the system can cross the barrier at a larger tip-surface separation, which results in both smaller hysteresis and smaller variation in the tip-surface force field.

As a further step, virtual nanomanipulation was also performed using tuning fork sensors and compared to simulations carried out with conventional silicon tips [Trevethan *et al.* (2007a)]. As a result, it could be established that the smaller oscillation amplitude of the tuning forks enhance the probability of manipulation events, which makes these sensors quite promising for future manipulation experiments.

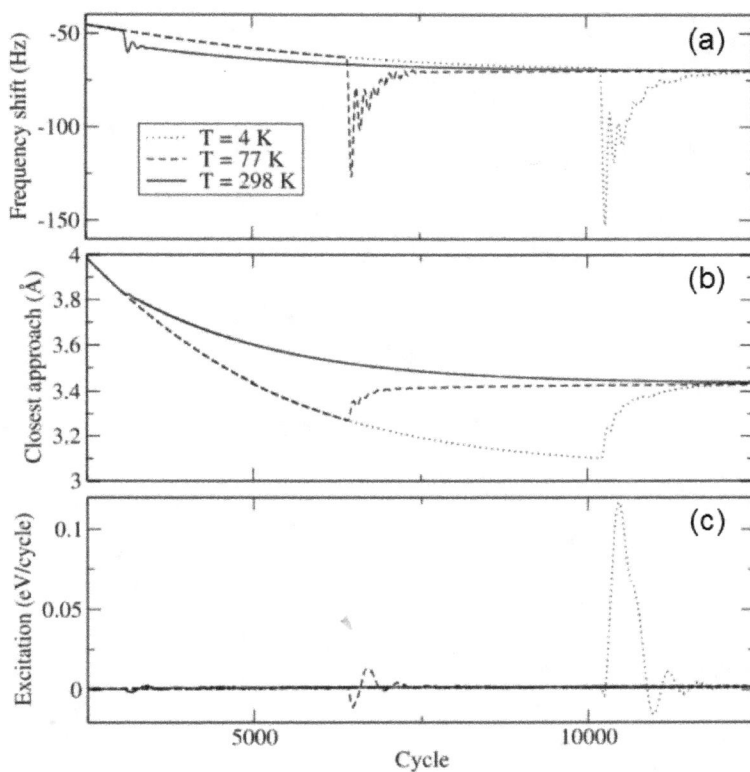

Fig. 13.5 Simulated (a) frequency shift, (b) closest approach, and (c) damping signals as a function of the cantilever oscillation cycle in NC-AFM manipulation of oxygen vacancies on a Mg(001) surface. Results at different temperatures (4, 77, and 298 K) are compared. Reprinted with permission from [Trevethan *et al.* (2007c)]. Copyright (2007) by the American Physical Society.

Bibliography

Agnoli, S., Castellarin-Cudia, C., Sambi, M., Surnev, S., Ramsey, M. G., Granozzi, G. and Netzer, F. P. (2003). Vanadium on $TiO_2(110)$: Adsorption site and sub-surface migration, *Surf. Sci.* **546**, p. 117.

Al-Abadleh, H. A. and Grassian, V. H. (2003). Oxide surfaces as environmental interfaces, *Surf. Sci. Rep.* **52**, pp. 63–161.

Altman, E. and Colton, R. (1992). Nucleation, growth, and structure of fullerene films on Au(111), *Surf. Sci.* **279**, p. 49.

Altman, E. and Colton, R. (1993). The interaction of C_{60} with noble-metal surfaces, *Surf. Sci.* **295**, p. 13.

Andersen, J. E. T. and Møller, P. J. (1991). Room-temperature interaction of ultrathin film yttrium with $SrTiO_3(100)$, $LaAlO_3(100)$, and $MgO(100)$ surfaces, *Phys. Rev. B* **44**, p. 13645.

Arce, P., Riera, G., Gorostiza, P. and Sanz, F. (2000). Atomic-layer expulsion in nanoindentations on an ionic single crystal, *Appl. Phys. Lett.* **77**, p. 839.

Ashcroft, N. W. and Mermin, N. D. (1976). *Solid State Physics* (Saunders College, Philadelphia).

Ashworth, T. V., Pang, C. L., Wincott, P. L., Vaughan, D. J. and Thornton, G. (2003). Imaging in situ cleaved $MgO(100)$ with non-contact atomic force microscopy, *Appl. Surf. Sci.* **210**, p. 2.

Avouris, P. and Wolkow, R. (1989). Scanning tunneling microscopy of insulators: CaF_2 epitaxy on Si(111), *Appl. Phys. Lett.* **55**, p. 1074.

Baker, J. and Lindgård, P. A. (1996). Monte Carlo determination of heteroepitaxial misfit structures, *Phys. Rev. B* **54**, p. R11137.

Bammerlin, M., Lüthi, R., Meyer, E., Baratoff, A., Lü, J., Guggisberg, M., Loppacher, C., Gerber, C. and Güntherodt, H. J. (1998). Dynamic SFM with true atomic resolution on alkali halide surfaces, *Appl. Phys. A* **66**, p. S293.

Baratoff, A., Gnecco, E., Roth, R., Steiner, P. and Meyer, E. (2009). Analytical solution of the athermal 1D Tomlinson model, *in preparation* .

Bartels, L., Meyer, G. and Rieder, K. H. (1999). The evolution of CO adsorption on Cu(111) as studied with bare and CO-funtionalized scanning tunneling tips, *Surf. Sci.* **432**, p. L621.

Barth, C. and Henry, C. R. (2003). Atomic resolution imaging of the (001) surface of UHV cleaved MgO by dynamic scanning force microscopy, *Phys. Rev. Lett.* **91**, p. 196102.

Barth, C. and Henry, C. R. (2004). High-resolution imaging of gold clusters on KBr(001) surfaces investigated by dynamic scanning force microscopy, *Nanotechnology* **15**, p. 1264.

Barth, C. and Henry, C. R. (2006a). Gold nanoclusters on alkali halide surfaces: Charging and tunneling, *Appl. Phys. Lett.* **89**, p. 252119.

Barth, C. and Henry, C. R. (2006b). Imaging nanoclusters in the constant height mode of the dynamic SFM, *Nanotechnology* **17**, p. S128.

Barth, C. and Henry, C. R. (2006c). Kelvin probe force microscopy on surfaces of UHV cleaved ionic crystals, *Nanotechnology* **17**, p. S155.

Barth, C. and Henry, C. R. (2007). Surface double layer on (001) surfaces of alkali halide crystals: A scanning force microscopy study, *Phys. Rev. Lett.* **98**, p. 136804.

Barth, C. and Henry, C. R. (2008). Imaging Suzuki precipitates on $NaCl:Mg^{2+}(001)$ by scanning force microscopy, *Phys. Rev. Lett.* **100**, p. 096101.

Barth, C. and Reichling, M. (2000). Resolving ions and vacancies at step edges on insulating surfaces, *Surf. Sci.* **470**, p. L99.

Barth, C. and Reichling, M. (2001). Imaging the atomic arrangements on the high-temperature reconstructed α-$Al_2O_3(0001)$ surface, *Nature* **414**, p. 54.

Batzill, M., Bardou, F. and Snowdon, K. J. (2001). Self-organization of large-area periodic nanowire arrays by glancing incidence ion bombardment of $CaF_2(111)$ surfaces, *J. Vac. Sci. Technol. A* **19**, p. 1829.

Batzill, M., Chaka, A. M. and Diebold, U. (2004). Surface oxygen chemistry of a gas-sensing material: $SnO_2(101)$, *Europhys. Lett.* **65**, p. 61.

Batzill, M. and Diebold, U. (2005). The surface and materials science of tin oxide, *Prog. Surf. Sci.* **79**, p. 47.

Batzill, M. and Diebold, U. (2007). Surface studies of gas sensing metal oxides, *Phys. Chem. Chem. Phys.* **9**, p. 2307.

Beck, K. M., Joly, A. G., Dupuis, N. F., Perozzo, P., Hess, W. P., Sushko, V. P. and Shluger, A. L. (2004). Laser control of product electronic state: Desorption from alkali halides, *J. Chem. Phys.* **120**, p. 2456.

Bennewitz, R., Barwich, V., Bammerlin, M., Loppacher, C., Guggisberg, M., Baratoff, A., Meyer, E. and Güntherodt, H. J. (1999). Ultrathin films of NaCl on Cu(111): A LEED and dynamic force microscopy study, *Surf. Sci.* **438**, p. 289.

Bennewitz, R., Foster, A. S., Kantorovich, L. N., Bammerlin, M., Loppacher, C., Schär, S., Guggisberg, M., Meyer, E. and Shluger, A. L. (2000). Atomically resolved edges and kinks of NaCl islands on Cu(111): Experiment and theory, *Phys. Rev. B* **62**, p. 2074.

Bennewitz, R., Pfeiffer, O., Schär, S., Barwich, V., Meyer, E. and Kantorovich, L. N. (2002). Atomic corrugation in nc-AFM of alkali halides, *Appl. Surf. Sci.* **188**, p. 232.

Bennewitz, R., Schär, S., Gnecco, E., Pfeiffer, O., Bammerlin, M. and Meyer, E. (2004). Atomic structure of alkali halide surfaces, *Appl. Phys. A* **78**, p. 837.

Bethge, H. (1964). Electron microscopic studies of surface structures and some relations to surface phenomena, *Surf. Sci.* **3**, p. 33.

Biersack, J. P. and Santner, E. (1982). Sputtering of alkali halides by 70 to 300 keV H, He, Ar ions, *Nucl. Instr. and Meth. in Phys. Res. B* **198**, p. 29.

Bikondoa, O., Pang, C. L., Muryn, C. A., Daniels, B. G., Ferrero, S., Michelangeli, E. and Thornton, G. (2004). Ordered overlayers of Ca on $TiO_2(110)$-1 × 1, *J. Phys. Chem. B* **108**, p. 16768.

Binnig, G., Quate, C. F. and Gerber, C. (1986). Atomic force microscope, *Phys. Rev. Lett.* **56**, p. 930.

Binnig, G., Rohrer, H., Gerber, C. and Weibel, E. (1982). Surface studies by scanning tunneling microscopy, *Phys. Rev. Lett.* **49**, p. 57.

Binnig, G., Rohrer, H., Gerber, C. and Weibel, E. (1983). 7×7 reconstruction on Si(111) resolved in real space, *Phys. Rev. Lett.* **50**, p. 120.

Bonnell, D. A. and Garra, J. (2008). Scanning probe microscopy of oxide surfaces: Atomic structure and properties, *Rep. Prog. Phys.* **71**, p. 044501.

Bradley, R. M. and Harper, J. M. E. (1988). Theory of ripple topography induced by ion-bombardment, *J. Vac. Sci. Technol. A* **6**, p. 2390.

Braun, K., Farias, D., Fölsch, S. and Rieder, K. (2000). Fractal growth of LiF on Ag(111) studied by low-temperature STM, *Surf. Sci.* **454**, p. 750.

Brinciotti, A., Piacentini, M. and Zema, N. (1994). Delayed emission in photon-stimulated desorption of alkali-halides, *Radiat. Eff. Defects Solids* **128**, p. 559.

Brinciotti, A., Zema, N. and Piacentini, M. (1991). Photostimulated desorption studies on potassium-iodide, *Radiat. Eff. Defects Solids* **119-121**, p. 559.

Bronshteyn, I. and Protsenko, N. (1970). Inelastic electron scattering and secondary electron emission of insulators, *Radio Eng. Electron. Phys.* **15**, p. 677.

Bruce, A. D. and Cowley, R. A. (1981). *Structural Phase Transitions* (Taylor and Francis, London).

Brune, H. (1998). Microscopic view of epitaxial metal growth: Nucleation and aggregation, *Surf. Sci. Rep.* **31**, p. 121.

Burke, S., Mativetsky, J., Fostner, S. and Grütter, P. (2007). C_{60} on alkali halides: Epitaxy and morphology studied by noncontact AFM, *Phys. Rev. B* **76**, p. 035419.

Burke, S. A., Mativetsky, J. M., Hoffmann, R. and Grütter, P. (2005). Nucleation and submonolayer growth of C_{60} on KBr, *Phys. Rev. Lett.* **94**, p. 096102.

Carpick, R. W., Ogletree, D. F. and Salmeron, M. (1997). Lateral stiffness: A new nanomechanical measurement for the determination of shear strengths with friction force microscopy, *Appl. Phys. Lett.* **70**, p. 1548.

Castell, M. (2002). Scanning tunneling microscopy of reconstructions on the $SrTiO_3(001)$ surface, *Surf. Sci.* **505**, pp. 1–13.

Chason, E., Mayer, T. M., Kellerman, B. K., McIlroy, D. T. and Howard, A. J. (1994). Roughening instability and evolution of the Ge(001) surface during ion sputtering, *Phys. Rev. Lett.* **72**, p. 3040.

Claessens, C. G., Gonzalez-Rodriguez, D. and Torres, T. (2002). Subphthalocyanines: Singular nonplanar aromatic compounds-synthesis, reactivity, and physical properties, *Chem. Rev.* **102**, p. 835.

de Graef, M. and McHenry, M. E. (2007). *Structure of Materials* (Cambridge University Press, Cambridge).

de Leeuw, N. H., Higgins, F. M. and Parker, S. C. (1999). Modeling the surface structure and stability of alpha-quartz, *J. Phys. Chem. B* **103**, p. 1270.

Degen, S., Krupski, A., Kralj, M., Langner, A., Becker, C., Sokolowski, M. and Wandelt, K. (2005). Determination of the coincidence lattice of an ultra thin Al_2O_3 film on $Ni_3Al(111)$, *Surf. Sci.* **576**, p. L57.

Diebold, U. (2003). The surface science of titanium dioxide, *Surf. Sci. Rep.* **48**, p. 53.

Diebold, U., Lehman, J., Mahmoud, T., Kuhn, M., Leonardelli, G., Hebenstreit, W., Schmid, M. and Varga, P. (1998). Intrinsic defects on a $TiO_2(1 \times 1)$ surface and their reaction with oxygen: A scanning tunneling microscopy study, *Surf. Sci.* **411**, pp. 137–153.

Ditchfield, R. and Seebauer, E. (1999). Direct measurement of ion-influenced surface diffusion, *Phys. Rev. Lett.* **82**, p. 1185.

Dulub, O., Diebold, U. and Kresse, G. (2003). Novel stabilization mechanism on polar surfaces: $ZnO(0001)$-Zn, *Phys. Rev. Lett.* **90**, p. 016102.

Dulub, O., Hebenstreit, W. and Diebold, U. (2000). Imaging cluster surfaces with atomic resolution: The strong metal-support interaction state of Pt supported on $TiO_2(110)$, *Phys. Rev. Lett.* **84**, p. 3646.

Dürig, U. (2000). Extracting interaction forces and complementary observables in dynamic probe microscopy, *Appl. Phys. Lett.* **76**, p. 1203.

Eby, J., Teegarden, K. and Dutton, D. (1959). Ultraviolet absorption of alkali halides, *Phys. Rev.* **116**, p. 1099.

Ejiri, A. (1987). Oblique-incidence absorption-spectroscopy studies of the plasmon in potassium halides, *Phys. Rev. B* **36**, p. 4946.

Elango, M. (1994). Hot holes in irradiated ionic solids, *Radiat. Eff. Defects Solids* **128**, p. 1.

Enevoldsen, G. H., Foster, A. S., Christensen, M. C., Lauritsen, J. V. and Besenbacher, F. (2007). Noncontact atomic force microscopy studies of vacancies and hydroxyls of $TiO_2(110)$: Experiments and atomistic simulations, *Phys. Rev. B* **76**, p. 205415.

Erlebacher, J., Aziz, M. J., Chason, E., Sinclair, M. B. and Floro, J. A. (1999). Spontaneous pattern formation on ion bombarded $Si(001)$, *Phys. Rev. Lett.* **82**, p. 2330.

Esch, F., Fabris, S., Zhou, L., Montini, T., Africh, C., Fornasiero, P., Comelli, G. and Rosei, R. (2005). Electron localization determines defect formation on ceria substrates, *Science* **309**, p. 752.

Facsko, S., Dekorsy, T., Koerdt, C., Trappe, C., Kurz, H., Vogt, A. and Hartnagel, H. L. (1999). Formation of ordered nanoscale semiconductor dots by ion sputtering, *Science* **285**, p. 1551.

Farias, D., Braun, K. F., Fölsch, S., Meyer, G. and Rieder, K. H. (2000). Observation of a novel nucleation mechanism at step edges: LiF molecules on $Ag(111)$, *Surf. Sci.* **470**, p. L93.

Fendrich, M., Kunstmann, T., Paulkowski, D. and Möller, R. (2007). Molecular resolution in dynamic force microscopy: Topography and dissipation for weakly interacting systems, *Nanotechnology* **18**, p. 084004.

Filleter, T., Maier, S. and Bennewitz, R. (2006). Atomic-scale yield and dislocation nucleation in KBr, *Phys. Rev. B* **73**, p. 155433.

Filleter, T., Paul, W. and Bennewitz, R. (2008). Atomic structure and friction of ultrathin films of KBr on Cu(100), *Phys. Rev. B* **77**, p. 035430.

Fölsch, S., Helms, A., Riemann, A., Repp, J., Meyer, G. and Rieder, K. H. (2002a). Nanoscale surface patterning by adsorbate-induced faceting and selective growth: NaCl on Cu(211), *Surf. Sci.* **497**, p. 113.

Fölsch, S., Helms, A., Zöphel, S., Repp, J., Meyer, G. and Rieder, K. H. (2000). Self-organized patterning of an insulator-on-metal system by surface faceting and selective growth: NaCl/Cu(211), *Phys. Rev. Lett.* **84**, p. 123.

Fölsch, S., Riemann, A., Repp, J., Meyer, G. and Rieder, K. H. (2002b). From atomic kinks to mesoscopic surface patterns: Ionic layers on vicinal metal surfaces, *Phys. Rev. B* **66**, p. 161409.

Forrest, S. R. (1997). Ultrathin organic films grown by organic molecular beam deposition and related techniques, *Chem. Rev.* **97**, p. 1793.

Foster, A. S., Barth, C., Shluger, A. L., Nieminen, R. M. and Reichling, M. (2002). Role of tip structure and surface relaxation in atomic resolution dynamic force microscopy: $CaF_2(111)$ as a reference surface, *Phys. Rev. B* **66**, p. 235417.

Foster, A. S., Barth, C., Shluger, A. L. and Reichling, M. (2001). Unambiguous interpretation of atomically resolved force microscopy images of an insulator, *Phys. Rev. Lett.* **86**, p. 2373.

Foster, A. S., Gal, A. Y., Gale, J. D., Lee, Y. J., Nieminen, R. M. and Shluger, A. L. (2004). Interaction of silicon dangling bonds with insulating surfaces, *Phys. Rev. Lett.* **92**, p. 036101.

Foster, A. S., Gal, A. Y., Nieminen, R. M. and Shluger, A. L. (2005). Probing organic layers on the $TiO_2(110)$ surface, *J. Phys. Chem. B* **109**, p. 4554.

French, T. M. and Somorjai, G. A. (1970). Composition and surface structure of the (0001) face of alumina by LEED, *J. Phys. Chem.* **74**, p. 2489.

Frenkel, J. (1946). *Kinetic Theory of Liquids* (Clarendon Press, Oxford).

Freund, H. J., Kuhlenbeck, H. and Staemmler, V. (1996). Oxide surfaces, *Rep. Prog. Phys.* **59**, p. 283.

Fu, Q. and Wagner, T. (2007). Interaction of nanostructured metal overlayers with oxide surfaces, *Surf. Sci. Rep.* **62**, pp. 431–498.

Fujii, S. and Fujihira, M. (2007). Atomic contrast on a point defect on $CaF_2(111)$ imaged by non-contact atomic force microscopy, *Nanotechnology* **18**, p. 084011.

Fukui, K., Namai, Y. and Iwasawa, Y. (2002). Imaging of surface oxygen atoms and their defect structures on $ceo_2(111)$ by noncontact atomic force microscopy, *Appl. Surf. Sci.* **188**, p. 252.

Fukui, K., Onishi, H. and Iwasawa, Y. (1997a). Atom-resolved image of the $TiO_2(110)$ surface by noncontact atomic force microscopy, *Phys. Rev. Lett.* **79**, p. 4202.

Fukui, K., Onishi, H. and Iwasawa, Y. (1997b). Imaging of individual formate ions adsorbed on $TiO_2(110)$ surface by non-contact atomic force microscopy, *Chem. Phys. Lett.* **280**, p. 296.

Fukui, K. and Sakai, M. (2006). Formation of one-dimensional C_{60} rows on $TiO_2(110)$-1×2-cross-link structure and their local polymerization, *J. Phys. Chem. B* **110**, pp. 21118–21123.

Fukuma, T., Kobayashi, K., Noda, K., Ishida, K., Horiuchi, T., Yamada, H. and Matsushiger, K. (2002). Molecular-scale non-contact AFM studies of ferroelectric organic thin films epitaxially grown on alkali halides, *Surf. Sci.* **516**, p. 103.

Gai, Z., Farnan, G., Pierce, J. and Shen, J. (2002a). Growth of low-dimensional magnetic nanostructures on an insulator, *Appl. Phys. Lett.* **81**, p. 742.

Gai, Z., Howe, J., Guo, J., Blom, D., Plummer, E. and Shen, J. (2005). Self-assembled FePt nanodot arrays with mono-dispersion and -orientation, *Appl. Phys. Lett.* **86**, p. 023107.

Gai, Z., Wu, B., Pierce, J., Farnan, G., Shu, D., Wang, M., Zhang, Z. and Shen, J. (2002b). Self-assembly of nanometer-scale magnetic dots with narrow size distributions on an insulating substrate, *Phys. Rev. Lett.* **89**, p. 235502.

Giessibl, F. (1995). Atomic resolution of the silicon (111)-(7×7) surface by atomic force microscopy, *Science* **267**, p. 68.

Giessibl, F. (1997). Forces and frequency shifts in atomic-resolution dynamic-force microscopy, *Phys. Rev. B* **56**, p. 16010.

Giessibl, F. J. (2003). Advances in atomic force microscopy, *Rev. Mod. Phys.* **75**, p. 949.

Giessibl, F. J., Hembacher, S., Herz, M., Schiller, C. and Mannhart, J. (2004). Stability considerations and implementation of cantilevers allowing dynamic force microscopy with optimal resolution: the qPlus sensor, *Nanotechnology* **15**, p. S79.

Glöckler, K., Sokolowski, M., Soukopp, A. and Umbach, E. (1996). Initial growth of insulating overlayers of NaCl on Ge(100) observed by scanning tunneling microscopy with atomic resolution, *Phys. Rev. B* **54**, p. 7705.

Gnecco, E., Bennewitz, R., Gyalog, T., Loppacher, C., Bammerlin, M., Meyer, E. and Güntherodt, H. J. (2000). Velocity dependence of atomic friction, *Phys. Rev. Lett.* **84**, p. 1172.

Gnecco, E., Bennewitz, R., Gyalog, T. and Meyer, E. (2001). Friction experiments on the nanometre scale, *J. Phys. Condens. Matter* **13**, p. R619.

Gnecco, E., Bennewitz, R. and Meyer, E. (2002). Abrasive wear on the atomic scale, *Phys. Rev. Lett.* **88**, p. 215501.

Gnecco, E. and Meyer, E. (eds.) (2007). *Fundamentals of Friction and Wear on the Nanoscale* (Springer, Berlin).

Gnecco, E., Socoliuc, A., Maier, S., Gessler, J., Glatzel, T., Baratoff, A. and Meyer, E. (2009). Dynamic superlubricity on insulating and conductive surfaces in ultra-high vacuum and ambient environment, *Nanotechnology* **20**, p. 025501.

Goryl, M., de Mongeot, F. B., Krok, F., Vevecka-Priftaj, A. and Szymonski, M. (2007). Leaky atomic traps: Upward diffusion of Au from nanoscale pits on ionic-crystal surfaces, *Phys. Rev. B* **76**, p. 075423.

Goryl, M., Such, B., Krok, F., Meisel, K., Kolodziej, J. J. and Szymonski, M. (2005). Atomic force microscopy studies of alkali halide surfaces nanostructured by DIET, *Surf. Sci.* **593**, p. 147.

Gratz, A. J., Manne, S. and Hansma, P. K. (1991). Atomic force microscopy of atomic-scale ledges and etch pits formed during dissolution of quartz, *Science* **251**, p. 1343.

Gritschneder, S., Degen, S., Becker, C., Wandelt, K. and Reichling, M. (2007). Atomic structure of a stripe phase on $Al_2O_3/Ni_3Al(111)$ revealed by scanning force microscopy, *Phys. Rev. B* **76**, p. 014123.

Guggisberg, M., Bammerlin, M., Loppacher, C., Pfeiffer, O., Abdurixit, A., Barwich, V., Bennewitz, R., Baratoff, A., Meyer, E. and Güntherodt, H. J. (2000). Separation of interactions by noncontact force microscopy, *Phys. Rev. B* **61**, p. 11151.

Gürgöze, M. (2005). On the representation of a cantilevered beam carrying a tip mass by an equivalent springmass system, *J. Sound Vib.* **202**, p. 538.

Haas, G., Menck, A., Brune, H., Barth, J., Venables, J. and Kern, K. (2000). Nucleation and growth of supported clusters at defect sites: Pd/MgO(001), *Phys. Rev. B* **61**, p. 11105.

Hamm, G., Barth, C., Becker, C., Wandelt, K. and Henry, C. R. (2006). Surface structure of an ultrathin alumina film on $Ni_3Al(111)$: A dynamic scanning force microscopy study, *Phys. Rev. Lett.* **97**, p. 126106.

Hammer, B. and Norskov, J. (1995). Why gold is the noblest of all the metals, *Nature* **376**, p. 238.

Hansen, H., Redinger, A., Messlinger, S., Stoian, G., Rosandi, Y., Urbassek, H., Linke, U. and Michely, T. (2006). Mechanisms of pattern formation in grazing-incidence ion bombardment of Pt(111), *Phys. Rev. B* **73**, p. 235414.

Hayashi, T., Yamashita, A., Maruno, T., Fölsch, S., Konami, H. and Hatano, M. (1995). Inplane ordering of a dibenzo[b,t]phthalocyaninato-Zn(ii) thin-film due to the atomic step arrays on a sapphire ($\bar{1}012$) surface, *J. Cryst. Growth* **156**, p. 245.

Hebenstreit, W., Redinger, J., Horozova, Z., Schmid, M., Podloucky, R. and Varga, P. (1999). Atomic resolution by STM on ultra-thin films of alkali halides: Experiment and local density calculations, *Surf. Sci.* **424**, pp. L321–L328.

Heemeier, M., Stempel, S., Shaikhutdinov, S. K., Libuda, J., Bumer, M., Oldman, R. J., Jackson, S. D. and Freund, H. J. (2003). On the thermal stability of metal particles supported on a thin alumina film, *Surf. Sci.* **523**, p. 103.

Heffelfinger, J. and Carter, C. (1997). Mechanisms of surface faceting and coarsening, *Surf. Sci.* **389**, p. 188.

Henrich, V. (1976). Thermal faceting of (110) and (111) surfaces of MgO, *Surf. Sci.* **57**, p. 385.

Henry, C. R. (1998). Surface studies of supported model catalysts, *Surf. Sci. Rep.* **31**, p. 231.

Henyk, M., Joly, A. G., Beck, K. M. and Hess, W. P. (2003). Photon stimulated desorption from KI: Laser control of I-atom velocity distributions, *Surf. Sci.* **528**, p. 219.

Herring, C. (1951). *The Physics of Powder Metallurgy* (McGraw-Hill, New York).

Hess, W. P., Joly, A. G., Beck, K. M., Henyk, M., Sushko, V. P., Trevisanutto, P. E. and Shluger, A. L. (2005). Laser control of desorption through selective surface excitation, *J. Phys. Chem. B* **109**, p. 19563.

Hess, W. P., Joly, A. G., Beck, K. M., Sushko, P. V. and Shluger, A. L. (2004). Determination of surface exciton energies by velocity resolved atomic desorption, *Surf. Sci.* **564**, p. 62.

Hess, W. P., Joly, A. G., Gerrity, D. P., Beck, K. M., Sushko, V. P. and Shluger, A. L. (2001). Selective laser desorption of ionic surfaces: Resonant surface excitation of KBr, *J. Chem. Phys.* **115**, p. 9463.

Hess, W. P., Joly, A. G., Gerrity, D. P., Beck, K. M., Sushko, V. P. and Shluger, A. L. (2002). Control of laser desorption using tunable single pulses and pulse pairs, *J. Chem. Phys.* **116**, p. 8144.

Heyde, M., Simon, G., Rust, H. and Freund, H. J. (2006). Probing adsorption sites on thin oxide films by dynamic force microscopy, *Appl. Phys. Lett.* **89**, p. 263107.

Hirth, S., Ostendorf, F. and Reichling, M. (2006). Lateral manipulation of atomic size defects on the $CaF_2(111)$ surface, *Nanotechnology* **17**, pp. S148–S154.

Hoeche, H., Toennies, J. P. and Vollmer, R. (1994). Combined electron-microscope surface-decoration and helium-atom-scattering study of the layer-by-layer photon-stimulated desorption from NaCl cleavage faces, *Phys. Rev. B* **50**, p. 679.

Hofer, W. A., Foster, A. S. and Shluger, A. L. (2003). Theories of scanning probe microscopes at the atomic scale, *Rev. Mod. Phys.* **75**, p. 1287.

Hoffmann, R., Baratoff, A., Hug, H., Hidber, H. R., v Löhneysen, H., and Güntherodt, H. J. (2007). Mechanical manifestations of rare atomic jumps in dynamic force microscopy, *Nanotechnology* **18**, p. 395503.

Hoffmann, R., Barth, C., Foster, A. S., Shluger, A. L., Hug, H. J., Güntherodt, H. J., Nieminen, R. M. and Reichling, M. (2005). Measuring site-specific cluster-surface bond formation, *J. Am. Chem. Soc.* **127**, p. 17863.

Hoffmann, R., Kantorovich, L. N., Baratoff, A., Hug, H. J. and Güntherodt, H. J. (2004). Sublattice identification in scanning force microscopy on alkali halide surfaces, *Phys. Rev. Lett.* **92**, p. 146103.

Hoffmann, R., Lantz, M., Hug, H. J., van Schendel, P. J. A., Kappenberger, P., Martin, S., Baratoff, A. and Güntherodt, H. J. (2002). Atomic resolution imaging and force versus distance measurements on KBr(001) using low temperature scanning force microscopy, *Appl. Surf. Sci.* **188**, p. 238.

Hojrup-Hansen, K., Ferrero, S. and Henry, C. (2004). Nucleation and growth kinetics of gold nanoparticles on MgO(100) studied by UHV-AFM, *Appl. Surf. Sci.* **226**, p. 167.

Hölscher, H., Ebeling, D. and Schwarz, U. (2008). Friction at atomic-scale surface steps: Experiment and theory, *Phys. Rev. Lett.* **101**, p. 246105.

Hooks, D. E., Fritz, T. and Ward, M. D. (2001). Epitaxy and molecular organization on solid substrates, *Adv. Mater.* **47**, p. 227.

Hoshino, A., Isoda, S., Kurata, H. and Kobayashi, T. (1994). Scanning tunneling microscope contrast of perylene-3,4,9,10-tetracarboxylic-dianhydride on graphite and its application to the study of epitaxy, *J. Appl. Phys.* **76**, p. 4113.

Israelachvili, J. (1991). *Intermolecular and Surface Forces* (Academic, London).

Itoh, N. (1976). Sputtering and dynamic interstitial motion in alkali halides, *Nucl. Instr. and Meth.* **132**, p. 201.

Itoh, N. and Stoneham, A. (2000). *Materials Modification by Electronic Excitation* (Cambridge Univers. Press, Cambridge).

Jammal, Y. A., Pooley, D. and Townsend, P. (1973). The role of exciton diffusion in the electron induced sputtering of alkali halides, *J. Phys. C: Solid State Phys.* **6**, p. 247.

Jiang, Q. and Zegenhagen, J. (1999). c(6×2) and c(4×2) reconstruction of SrTiO$_3$(001), *Surf. Sci.* **425**, p. 343.

Kawasaki, M. (2003). Growth-induced inhomogeneities in synthetic quartz crystals revealed by the cathodoluminescence method, *J. Cryst. Growth* **247**, p. 185.

Kelly, R. (1979). Thermal effects in sputtering, *Surf. Sci.* **90**, p. 280.

Kelly, R. and Lam, N. (1973). The sputtering of oxides part I: a survey of the experimental results, *Radiat. Eff. Defects Solids* **19**, p. 39.

Kitahara, T., Sugawara, A., Sano, H. and Mizutani, G. (2003a). Anisotropic optical second-harmonic generation from the Au nanowire array on the NaCl(110) template, *Appl. Surf. Sci.* **219**, p. 271.

Kitahara, T., Sugawara, A., Sano, H. and Mizutani, G. (2003b). Optical second-harmonic spectroscopy of Au nanowires, *J. Appl. Phys.* **219**, p. 271.

Klust, A., Ohta, T., Bostwick, A., Rotenberg, E., Yu, Q., Ohuchi, F. and Olmstead, M. (2005). Electronic structure evolution during the growth of ultrathin insulator films on semiconductors: From interface formation to bulklike CaF$_2$/Si(111) films, *Phys. Rev. B* **72**, p. 205336.

Klust, A., Ohta, T., Bostwick, A., Yu, Q., Ohuchi, F. and Olmstead, M. (2004). Atomically resolved imaging of a CaF bilayer on Si(111): Subsurface atoms and the image contrast in scanning force microscopy, *Phys. Rev. B* **69**, p. 035405.

Klust, A., Pietsch, H. and Wollschläger, J. (1998). Growth of CaF$_2$ on Si(111): Imaging of the CaF interface by friction force microscopy, *Appl. Phys. Lett.* **73**, p. 1967.

Kolodziej, J., Postawa, Z., Czuba, P., Piatkowsi, P. and Szymonski, M. (1994). A comparison of electron-stimulated desorption of halogen atoms from different alkali-halide single-crystals, *Radiat. Eff. Defects Solids* **128**, p. 47.

Kolodziej, J. J., Such, B., Czuba, P., Krok, F., Piatkowski, P., Struski, P., Szymonski, M., Bennewitz, R., Schär, S. and Meyer, E. (2001). Frenkel defect interactions at surfaces of irradiated alkali halides studied by non-contact atomic-force microscopy, *Surf. Sci.* **482**, p. 903.

Kolodziej, J. J., Such, B., Czuba, P., Krok, F., Piatkowski, P. and Szymonski, M. (2002). Scanning-tunneling/atomic-force microscopy study of the growth of KBr films on InSb(001), *Surf. Sci.* **506**, p. 12.

Kresse, G., Schmid, M., Napetschnig, E., Shishkin, M., Köhler, L. and Varga, P. (2005). Structure of the ultrathin aluminum oxide film on NiAl(110), *Science* **308**, p. 1440.

Krok, F., Kolodziej, J. J., Such, B., Czuba, P., Piatkowski, P., Struski, P. and Szymonski, M. (2004a). Desorption and surface topography changes induced by He^+ ion bombardment of alkali halides, *Nucl. Instr. and Meth. in Phys. Res. B* **226**, p. 601.

Krok, F., Kolodziej, J. J., Such, B., Czuba, P., Struski, P., Piatkowski, P. and Szymonski, M. (2004b). Dynamic force microscopy and kelvin probe force microscopy of KBr film on InSb(001) surface at submonolayer coverage, *Surf. Sci.* **566**, p. 63.

Kroto, H. W., Heath, J. R., O'Brien, S. C., Curl, R. F. and Smalley, R. E. (1985). C_{60}: Buckminsterfullerene, *Nature* **318**, p. 162.

Krylov, S. Y., Dijksman, J. A., van Loo, W. A. and Frenken, J. W. M. (2006). Stick-slip motion in spite of a slippery contact: Do we get what we see in atomic friction? *Phys. Rev. Lett.* **97**, p. 166106.

Krylov, S. Y., Jinesh, K. B., Valk, H., Dienwiebel, M. and Frenken, J. W. M. (2005). Thermally induced suppression of friction at the atomic scale, *Phys. Rev. E* **71**, p. 065101.

Kubo, T. and Nozoye, H. (2001). Surface structure of $SrTiO_3(100)$-$(\sqrt{5} \times \sqrt{5})$-R26.6°, *Phys. Rev. Lett.* **86**, p. 1801.

Kubo, T., Okano, A., Kanasaki, J., Ishikawa, K., Nakai, Y. and Itoh, N. (1994). Emission of Na atoms from undamaged and slightly damaged NaCl (100) surfaces by electronic excitation, *Phys. Rev. B* **49**, p. 4931.

Kulawik, M., Nilius, N. and Freund, H. J. (2006). Influence of the metal substrate on the adsorption properties of thin oxide layers: Au atoms on a thin alumina film on NiAl(110), *Phys. Rev. Lett.* **96**, p. 036103.

Kunstmann, T., Schlarb, A., Fendrich, M., Wagner, T., Möller, R. and Hoffmann, R. (2005). Dynamic force microscopy study of 3,4,9,10-perylenetetracarboxylic dianhydride on KBr(001), *Phys. Rev. B* **71**, p. 121403.

Lai, X., Clair, T. P. S., Valden, M. and Goodman, D. W. (1998). Scanning tunneling microscopy studies of metal clusters supported on $TiO_2(110)$: Morphology and electronic structure, *Prog. Surf. Sci.* **59**, p. 25.

Lantz, M. A., Hoffmann, R., Foster, A. S., Baratoff, A., Hug, H. J., Hidber, H. R. and Güntherodt, H. J. (2006). Site-specific force-distance characteristics on NaCl(001): Measurements versus atomistic simulations, *Phys. Rev. B* **74**, p. 245426.

Lauritsen, J. V., Foster, A. S., Olesen, G. H., Christensen, M. C., Kühnle, A., Helveg, S., Rostrup-Nielsen, J., Clausen, B., Reichling, M. and Besenbacher, F. (2006). Chemical identification of point defects and adsorbates on a metal oxide surface by atomic force microscopy, *Nanotechnology* **17**, p. 3436.

Li, X., Beck, R. D. and Whetten, R. L. (1992). Photon-stimulated ejection of atoms from alkali-halide nanocrystals, *Phys. Rev. Lett.* **68**, p. 3420.

Liljeroth, P., Repp, J. and Meyer, G. (2007). Current-induced hydrogen tautomerization and conductance switching of naphthalocyanine molecules, *Science* **317**, p. 1203.

Lin, J., Petrovykh, D., Kirakosian, A., Rauscher, H., Himpsel, F. and Dowben, P. (2001). Self-assembled Fe nanowires using organometallic chemical vapor deposition and CaF$_2$ masks on stepped Si(111), *Appl. Phys. Lett.* **78**, p. 829.

Lin, W., Kuo, C., Luo, M., Song, K. and Lin, M. (2005). Self-aligned Co nanoparticle chains supported by single-crystalline Al$_2$O$_3$/NiAl(100) template, *Appl. Phys. Lett.* **86**, p. 043105.

Lin, W., Wong, S., Huang, P., Wu, C., Xu, B., Chiang, C., Yen, H. and Lin, M. (2006). Controlled growth of Co nanoparticle assembly on nanostructured template Al$_2$O$_3$/NiAl(100), *Appl. Phys. Lett.* **89**, p. 153111.

Liu, D., Albridge, R. G., Barnes, A. V., Bunton, P. H., Ewig, C. S., Tolk, N. H. and Szymonski, M. (1993). The desorption of excited atoms by photon bombardment of alkali-halide crystals, *Surf. Sci.* **281**, p. L353.

Livshits, A. I. and Shluger, A. L. (1997). Model of noncontact scanning force microscopy on ionic surfaces, *Phys. Rev. B* **56**, p. 12482.

Livshits, A. I., Shluger, A. L., Rohl, A. L. and Foster, A. S. (1999). Model of noncontact scanning force microscopy on ionic surfaces, *Phys. Rev. B* **59**, p. 2436.

Loppacher, C., Guggisberg, M., Pfeiffer, O., Meyer, E., Bammerlin, M., Lüthi, R., Schlittler, R., Gimzewski, J. K., Tang, H. and Joachim, C. (2003). Direct determination of the energy required to operate a single molecule switch, *Phys. Rev. Lett.* **90**, p. 066107.

Loppacher, C., Zerweck, U. and Eng, L. (2004). Kelvin probe force microscopy of alkali chloride thin films on Au(111), *Nanotechnology* **15**, p. S9.

Loppacher, C., Zerweck, U., Eng, L., Gemming, S., Seifert, G., Olbrich, C., Morawetz, K. and Schreiber, M. (2006). Adsorption of PTCDA on a partially KBr covered Ag(111) substrate, *Nanotechnology* **17**, p. 1568.

Luo, M., Chiang, C., Shiu, H., Sartale, S. and Kuo, C. (2006a). Patterning Co nanoclusters on thin-film Al$_2$O$_3$/NiAl(100), *Nanotechnology* **17**, p. 360.

Luo, M., Chiang, C., Shiu, H., Sartale, S., Wang, T., Chen, P. and Kuo, C. (2006b). Growth of Co clusters on thin films Al$_2$O$_3$/NiAl(100), *J. Chem. Phys.* **124**, p. 164709.

Luo, M. F., Shiu, H. W., Ten, M. H., Sartale, S. D., Chiang, C. I., Lin, Y. C. and Hsu, Y. J. (2008). Growth and electronic properties of au nanoclusters on thin-film Al$_2$O$_3$/NiAl(100) studied by scanning tunnelling microscopy and photoelectron spectroscopy with synchrotron radiation, *Surf. Sci.* **602**, p. 241.

Lüth, H. (1996). *Surfaces and Interfaces of Solid Materials* (Springer, Berlin).

Lüthi, R., Meyer, E., Bammerlin, M., Howald, L., Haefke, H., Lehmann, T., Loppacher, C., Güntherodt, H. J., Gyalog, T. and Thomas, H. (1996). Friction on the atomic scale: An ultrahigh vacuum atomic force microscopy study on ionic crystals, *J. Vac. Sci. Technol. B* **14**, p. 1280.

Lüthi, R., Meyer, E., Haefke, H., Howald, L., Gutmannsbauer, W., Guggisberg, M., Bammerlin, M. and Güntherodt, H. J. (1995). Nanotribology: an UHV-SFM study on thin films of C$_{60}$ and AgBr, *Surf. Sci.* **338**, p. 247.

Lüthi, R., Meyer, E., Haefke, H., Howald, L., Gutmannsbauer, W. and Güntherodt, H. J. (1994). Sled-type motion on the nanometer scale: Determination of dissipation and cohesive energy of C_{60}, *Science* **266**, p. 1979.

MacLaren, S. W., Baker, J. E., Finnegan, N. L. and Loxton, C. M. (1992). Surface-roughness development during sputtering of GaAs and InP - evidence for the role of surface-diffusion in ripple formation and sputter cone development, *J. Vac. Sci. Technol. A* **10**, p. 468.

Maier, S., Fendt, L. A., Zimmerli, L., Glatzel, T., Pfeiffer, O., Diederich, F. and Meyer, E. (2008a). Nanoscale engineering of molecular porphyrin wires on insulating surfaces, *Small* **4**, p. 1115.

Maier, S., Gnecco, E., Baratoff, A., Bennewitz, R. and Meyer, E. (2008b). Atomic-scale friction modulated by a buried interface: Combined atomic and friction force microscopy experiments, *Phys. Rev. B* **78**, p. 045432.

Maier, S., Pfeiffer, O., Glatzel, T., Meyer, E., Filleter, T. and Bennewitz, R. (2007). Asymmetry in the reciprocal epitaxy of NaCl and KBr, *Phys. Rev. B* **75**, p. 195408.

Maier, S., Sang, Y., Filleter, T., Grant, M., Bennewitz, R., Gnecco, E. and Meyer, E. (2005). Fluctuations and jump dynamics in atomic friction experiments, *Phys. Rev. B* **72**, p. 245418.

Makeev, M. A. and Barabasi, A. L. (1997). Ion-induced effective surface diffusion in ion sputtering, *Appl. Phys. Lett.* **71**, p. 2800.

Makeev, M. A., Cuerno, R. and Barabasi, A. L. (2002). Morphology of ion-sputtered surfaces, *Nucl. Instr. and Meth. in Phys. Res. B* **197**, p. 185.

Mate, C., McClelland, G., Erlandsson, R. and Chiang, S. (1987). Atomic-scale friction of a tungsten tip on a graphite surface, *Phys. Rev. Lett.* **59**, p. 1942.

Mativetsky, J., Burke, S., Fostner, S. and Grutter, P. (2007). Templated growth of 3,4,9,10-perylenetetracarboxylic dianhydride molecules on a nanostructured insulator, *Nanotechnology* **18**, p. 105303.

Mativetsky, J. M., Miyahara, Y., Fostner, S., Burke, S. A. and Grutter, P. (2006). Use of an electron-beam evaporator for the creation of nanostructured pits in an insulating surface, *Appl. Phys. Lett.* **88**, p. 233121.

Matthey, D., Wang, J. G., Wendt, S., Matthiesen, J., Schaub, R., Laegsgaard, E., Hammer, B. and Besenbacher, F. (2007). Enhanced bonding of gold nanoparticles on oxidized $TiO_2(110)$, *Science* **315**, p. 1692.

Meyer, E., Hug, H. and Bennewitz, R. (2004). *Scanning Probe Microscopy: The Lab on a Tip* (Springer, Berlin).

Meyer, E., Lüthi, R., Howald, L., Bammerlin, M., Guggisberg, M. and Güntherodt, H. J. (1996). Friction on the atomic scale: An ultrahigh vacuum atomic force microscopy study on ionic crystals, *J. Vac. Sci. Technol. B* **14**, p. 1285.

Moors, M., Krupski, A., Degen, S., Kralj, M., Becker, C. and Wandelt, K. (2008). Scanning tunneling microscopy and spectroscopy investigations of copper phthalocyanine adsorbed on $Al_2O_3/Ni_3Al(111)$, *Appl. Surf. Sci.* **254**, p. 4251.

Moriarty, P. (2001). Nanostructured materials, *Rep. Prog. Phys.* **64**, p. 297.

Mullins, W. (1957). Theory of thermal grooving, *J. Appl. Phys.* **28**, p. 333.

Mussi, V., Granone, F., Boragno, C., de Mongeot, F. B., Valbusa, U., Marolo, T. and Montereali, R. M. (2006). Surface nanostructuring and optical activation of lithium fluoride crystals by ion beam irradiation, *Appl. Phys. Lett.* **88**, p. 103116.

Nilius, N., Wallis, T. M. and Ho, W. (2003). Influence of a heterogeneous Al_2O_3 surface on the electronic properties of single Pd atoms, *Phys. Rev. Lett.* **90**, p. 046808.

Nishi, R., andY . Seino, D. M., Yi, I. and Morita, S. (2006). Non-contact atomic force microscopy study of atomic manipulation on an insulator surface by nanoindentation, *Nanotechnology* **17**, p. S142.

Nony, L., Bennewitz, R., Pfeiffer, O., Gnecco, E., Baratoff, A., Meyer, E., Eguchi, T., Gourdon, A. and Joachim, C. (2004a). Cu-TBPP and PTCDA molecules on insulating surfaces studied by ultra-high-vacuum non-contact AFM, *Nanotechnology* **15**, p. S91.

Nony, L., Gnecco, E., Baratoff, A., Alkauskas, A., Bennewitz, R., Pfeiffer, O., Maier, S., Wetzel, A., Meyer, E. and Gerber, C. (2004b). Observation of individual molecules trapped on a nanostructured insulator, *Nano Lett.* **4**, p. 2185.

Nörenberg, H. and Briggs, G. (1997). Defect structure of nonstoichiometric ceo2(111) surfaces studied by scanning tunneling microscopy, *Phys. Rev. Lett.* **79**, p. 4222.

Norskov, J. K. (1990). Chemisorption on metal surfaces, *Rep. Prog. Phys.* **53**, p. 1253.

Ogata, S., Li, J., Hirosaki, N., Shibutani, H. and Yip, S. (2004). Ideal shear strain of metals and ceramics, *Phys. Rev. B* **70**, p. 104104.

Olsson, F. E., Persson, M., Repp, J. and Meyer, G. (2005). Scanning tunneling microscopy and spectroscopy of NaCl overlayers on the stepped Cu(311) surface: Experimental and theoretical study, *Phys. Rev. B* **71**, p. 075419.

Onishi, H., Fukui, K. and Iwasawa, Y. (1999). Space-correlation analysis of formate ions adsorbed on $TiO_2(110)$, *Jap. J. Appl. Phys.* **38**, p. 3830.

Onishi, H., Sasahara, A., Uetsuka, H. and Ishibashi, T. (2002). Molecule-dependent topography determined by noncontact atomic force microscopy: carboxylates on $TiO_2(110)$, *Appl. Surf. Sci.* **188**, p. 257.

Ostendorf, F., Torbrügge, S. and Reichling, M. (2008). Atomic scale evidence for faceting stabilization of a polar oxide surface, *Phys. Rev. B* **77**, p. 041405.

Oura, K., Lifshits, V. G., Saranin, A. A., Zotov, A. V. and Katayama, M. (2003). *Surface Science: An Introduction* (Springer, Berlin).

Overeijnder, H., Haring, A. and de Vries, A. E. (1978a). Sputtering processes during 6 keV Xe ion-beam bombardment of halides, *Radiat. Eff. Defects Solids* **36**, p. 189.

Overeijnder, H., Haring, A. and de Vries, A. E. (1978b). Sputtering processes of alkali-halides during 6 keV Xe^+ ion-bombardment, *Radiat. Eff. Defects Solids* **37**, p. 205.

Overeijnder, H., Szymonski, M., Haring, A. and de Vries, A. E. (1977). The mechanism of the electron sputtering process of alkali halides, *Phys. Status Solidi* **81**, p. K11.

Overeijnder, H., Szymonski, M., Haring, A. and de Vries, A. E. (1978c). Electron sputtering of alkali-halides - study of its dependence on beam energy and target temperature, *Radiat. Eff. Defects Solids* **38**, p. 21.

Pai, W. W., Hsu, C. L., Lin, M. C., Lin, K. C. and Tang, T. B. (2004). Structural relaxation of adlayers in the presence of adsorbate-induced reconstruction: C_{60}/Cu(111), *Phys. Rev. B* **69**, p. 125405.

Pakarinen, O. H., Barth, C., Foster, A. S. and Henry, C. R. (2008). Imaging the real shape of nanoclusters in scanning force microscopy, *J. Appl. Phys.* **103**, p. 0543413.

Pakarinen, O. H., Barth, C., Foster, A. S., Nieminen, R. M. and Henry, C. R. (2006). High-resolution scanning force microscopy of gold nanoclusters on the KBr(001) surface, *Phys. Rev. B* **73**, p. 235428.

Pang, C. L., Muryn, C. A., Woodhead, A. P., Raza, H., Haycock, S. A., Dhanak, V. and Thornton, G. (2005). Low-coverage condensation of K on $TiO_2(110)1\times1$, *Surf. Sci.* **583**, p. L147.

Pang, C. L., Raza, H., Haycock, S. A. and Thornton, G. (2000). Growth of copper and palladium on α-$Al_2O_3(0001)$, *Surf. Sci.* **460**, p. L510.

Peimann, C. and Skibowski, M. (1971). Dielectric properties of rubidium halide crystals in extreme ultraviolet up to 30 eV, *Phys. Status Solidi B* **46**, p. 655.

Pfeiffer, O., Gnecco, E., Zimmerli, L., Maier, S., Meyer, E., Nony, L., Bennewitz, R., Diederich, F., Fang, H. and Bonifazi, D. (2007). Force microscopy on insulators: Imaging of organic molecules, *J. Phys. Confer. Series* **61**, p. 1357.

Pivetta, M., Patthey, F., Stengel, M., Baldereschi, A. and Schneider, W. D. (2005). Local work function Moiré pattern on ultrathin ionic films: NaCl on Ag(100), *Phys. Rev. B* **72**, p. 115404.

Postawa, Z. and Szymonski, M. (1989). Directional emission of nonthermal halogen atoms by electron-bombardment of alkali-halides, *Phys. Rev. B* **39**, p. 12950.

Puchin, V., Shluger, A. and Itoh, N. (1993). Theoretical studies of atomic emission and defect formation by electronic excitation at the (100) surface of NaCl, *Phys. Rev. B* **47**, p. 10760.

Puchin, V., Shluger, A., Nakai, Y. and Itoh, N. (1994). Theoretical-study of Na-atom emission from NaCl(100) surfaces, *Phys. Rev. B* **49**, p. 11364.

Qiu, X., Nazin, G. and Ho, W. (2004). Vibronic states in single molecule electron transport, *Phys. Rev. Lett.* **92**, p. 206102.

Ramoino, L., von Arx, M., Schintke, S., Baratoff, A., Güntherodt, H. J. and Jung, T. A. (2006). Layer-selective epitaxial self-assembly of porphyrins on ultrathin insulators, *Chem. Phys. Lett.* **417**, p. 22.

Rauscher, H., Jung, T., Lin, J., Kirakosian, A., Himpsel, F., Rohr, U. and Müllen, K. (1999). One-dimensional confinement of organic molecules via selective adsorption on CaF_1 versus CaF_2, *Chem. Phys. Lett.* **303**, p. 363.

Reichling, M. and Barth, C. (1999). Scanning force imaging of atomic size defects on the $CaF_2(111)$ surface, *Phys. Rev. Lett.* **83**, p. 768.

Reichling, M., Huisinga, M., Gogoll, S. and Barth, C. (1999). Degradation of the $caf_2(111)$ surface by air exposure, *Surf. Sci.* **439**, p. 181.

Repp, J., Fölsch, S., Meyer, G. and Rieder, K. H. (2001). Ionic films on vicinal metal surfaces: Enhanced binding due to charge modulation, *Phys. Rev. Lett.* **86**, p. 252.

Repp, J., Meyer, G., Olsson, F. and Persson, M. (2004a). Controlling the charge state of individual gold adatoms, *Science* **305**, p. 493.

Repp, J., Meyer, G., Paavilainen, S., Olsson, F. and Persson, M. (2005a). Scanning tunneling spectroscopy of Cl vacancies in NaCl films: Strong electron-phonon coupling in double-barrier tunneling junctions, *Phys. Rev. Lett.* **95**, p. 225503.

Repp, J., Meyer, G. and Rieder, K. H. (2004b). Snell's law for surface electrons: Refraction of an electron gas imaged in real space, *Phys. Rev. Lett.* **92**, p. 036803.

Repp, J., Meyer, G., Stojkovic, S., Gourdon, A. and Joachim, C. (2005b). Molecules on insulating films: Scanning-tunneling microscopy imaging of individual molecular orbitals, *Phys. Rev. Lett.* **94**, p. 026803.

Riedo, E., Gnecco, E., Bennewitz, R., Meyer, E. and Brune, H. (2003). Interaction potential and hopping dynamics governing sliding friction, *Phys. Rev. Lett.* **91**, p. 084502.

Rignanese, G. M., Vita, A. D., Charlier, J. C., Gonze, X. and Car, R. (2000). First-principles molecular-dynamics study of the (0001) α-quartz surface, *Phys. Rev. B* **61**, p. 13250.

Rosenhahn, A., Schneider, J., Becker, C. and Wandelt, K. (2000). Oxidation of $Ni_3Al(111)$ at 600, 800, and 1050 k investigated by scanning tunneling microscopy, *J. Vac. Sci. Technol. A* **18**, p. 1923.

Rossnagel, S., Robinson, R. and Kaufman, H. (1982). Impact enhanced surface-diffusion during impurity induced sputter cone formation, *Surf. Sci.* **123**, p. 89.

Saeed, S., Sinha, O., Krok, F. and Szymonski, M. (2008). Nanometer-scale patterning of alkali halide surfaces by ion bombardment, *Appl. Surf. Sci.* .

Salminen, O., Riihola, P., Ozols, A. and Viitala, T. (1996). Spatially resolved optical studies of F-center diffusion in KBr crystals, *Phys. Rev. B* **53**, p. 6129.

Sartale, S., Lin, K., Chiang, C., Luo, M. and Kuo, C. (2006). Engineering patterns of Co nanoclusters on thin film $Al_2O_3/NiAl(100)$ using scanning tunneling microscopy manipulation techniques, *Appl. Phys. Lett.* **89**, p. 0613118.

Sasahara, A., Uetsuka, H., Ishibashi, T. and Onishi, H. (2002). A needle-like organic molecule imaged by noncontact atomic force microscopy, *Appl. Surf. Sci.* **188**, p. 265.

Sasahara, A., Uetsuka, H. and Onishi, H. (2001a). Image topography of alkyl-substituted carboxylates observed by noncontact atomic force microscopy, *Surf. Sci.* **481**, p. L437.

Sasahara, A., Uetsuka, H. and Onishi, H. (2001b). Noncontact atomic force microscope topography dependent on the electrostatic dipole field of individual molecules, *Phys. Rev. B* **64**, p. 121406.

Sasahara, A., Uetsuka, H. and Onishi, H. (2001c). Single-molecule analysis by noncontact atomic force microscopy, *J. Phys. Chem. B* **105**, p. 1.

Sasahara, A., Uetsuka, H. and Onishi, H. (2003). Chemical identification of carboxylate surfactants with one-fluorine-atom sensitivity achieved by noncontact atomic force microscopy, *Langmuir* **19**, p. 7474.

Schintke, S., Messerli, S., Pivetta, M., Patthey, F., Libioulle, L., Stengel, M., Vita, A. D. and Schneider, W. D. (2001). Insulator at the ultrathin limit: Mgo on Ag(001), *Phys. Rev. Lett.* **87**, p. 276801.

Schirmeisen, A., Weiner, D. and Fuchs, H. (2006). Single-atom contact mechanics: From atomic scale energy barrier to mechanical relaxation hysteresis, *Phys. Rev. Lett.* **97**, p. 136101.

Sigmund, P. (1969). Theory of sputtering. I. sputtering yield of amorphous and polycrystalline targets, *Phys. Rev.* **184**, p. 383.

Sigmund, P. (1973). Mechanism of surface micro-roughening by ion-bombardment, *J. Mater. Res.* **8**, p. 1545.

Sigmund, P. and Szymonski, M. (1984). Temperature-dependent sputtering of metals and insulators, *Appl. Phys. A* **33**, p. 141.

Silly, F. and Castell, M. R. (2005). Selecting the shape of supported metal nanocrystals: Pd huts, hexagons, or pyramids on $SrTiO_3(001)$, *Phys. Rev. Lett.* **94**, p. 046103.

Silly, F. and Castell, M. R. (2006). Bimodal growth of au on $SrTiO_3(001)$, *Phys. Rev. Lett.* **96**, p. 086104.

Silly, F., Powell, A. C., Martin, M. G. and Castell, M. R. (2005). Growth shapes of supported Pd nanocrystals on $SrTiO_3(001)$, *Phys. Rev. B* **72**, p. 165403.

Smith, K. M. (1975). *Porhyrins and Metalloporphyrins* (Elsevier, New York).

Socoliuc, A., Bennewitz, R., Gnecco, E. and Meyer, E. (2004). Transition from stick-slip to continuous sliding in atomic friction: Entering a new regime of ultralow friction, *Phys. Rev. Lett.* **92**, p. 134301.

Socoliuc, A., Gnecco, E., Bennewitz, R. and Meyer, E. (2003). Ripple formation induced in localized abrasion, *Phys. Rev. B* **68**, p. 115416.

Socoliuc, A., Gnecco, E., Maier, S., Pfeiffer, O., Baratoff, A., Bennewitz, R. and Meyer, E. (2006). Atomic-scale control of friction by actuation of nanometer-sized contacts, *Science* **313**, p. 207.

Song, K. and Williams, R. (1993). *Self-Trapped Excitons* (Springer, Berlin).

Sterrer, M., Risse, T., Giordano, L., Heyde, M., Nilius, N., Rust, H., Pacchioni, G. and Freund, H. (2007). Palladium monomers, dimers, and trimers on the MgO(001) surface viewed individually, *Angew. Chem. Int. Ed.* **46**, p. 8703.

Sterrer, M., Yulikov, M., Fischbach, E., Heyde, M., Rust, H. P., Pacchioni, G., Risse, T. and Freund, H. J. (2006). Interaction of gold clusters with color centers on MgO(001) films, *Angew. Chem. Int. Ed.* **45**, p. 2630.

Steurer, W., Apfolter, A., Koch, M., Sarlat, T., Søndergård, E., Ernst, W. and Holst, B. (2007). The structure of the alpha-quartz (0001) surface investigated using helium atom scattering and atomic force microscopy, *Surf. Sci.* **601**, p. 4407.

Such, B., Czuba, P., Piatkowski, P. and Szymonski, M. (2000a). AFM studies of electron-stimulated desorption process of KBr(001) surface, *Surf. Sci.* **451**, p. 203.

Such, B., Kolodziej, J., Czuba, P., Piatkowski, P., Struski, P., Krok, F. and Szymonski, M. (2000b). Surface topography dependent desorption of alkali halides, *Phys. Rev. Lett.* **85**, p. 2621.

Sugawara, A., Coyle, T., Hembree, G. G. and Scheinfein, M. R. (1997a). Self-organized Fe nanowire arrays prepared by shadow deposition on NaCl(110) templates, *Appl. Phys. Lett.* **70**, p. 1043.

Sugawara, A., Hembree, G. G. and Scheinfein, M. R. (1997b). Self-organized mesoscopic magnetic structures, *J. Appl. Phys.* **82**, p. 5662.

Sugimoto, Y., Abe, M., Hirayama, S., Oyabu, N., Custance, O. and Morita, S. (2005). Atom inlays performed at room temperature using atomic force microscopy, *Nat. Mater.* **4**, p. S156.

Sushko, M. L., Gal, A. Y., Watkins, M. and Shluger, A. L. (2006). Modelling of non-contact atomic force microscopy imaging of individual molecules on oxide surfaces, *Nanotechnology* **17**, p. 2062.

Suzuki, K. (1955). X-ray studies on the structures of solid solutions $NaCl$-$CaCl_2$ II. structures of {111} and {310} plate-zones, *J. Phys. Soc. Jpn.* **10**, p. 794.

Swank, R. K. and Brown, F. C. (1963). Lifetime of the excited F center, *Phys. Rev. B* **130**, p. 34.

Szymonski, M. (1980). On the model of the electron sputtering process of alkali halides, *Radiat. Eff. Defects Solids* **52**, p. 9.

Szymonski, M. (1982). Sputtering mechanisms of compound solids, *Nucl. Instr. and Meth. in Phys. Res. B* **194**, p. 523.

Szymonski, M. (1993). Electronic sputtering of alkali halides, *Kgl. Dan. Vid. Selsk. Mat.-Fys. Medd.* **43**, p. 495.

Szymonski, M., A. Droba and, M. G., Kolodziej, J. J. and Krok, F. (2006). Alkali halide decomposition and desorption by photons - the role of excited point defects and surface topographies, *J. Phys. Condens. Matter* **18**, p. S1547.

Szymonski, M. and de Vries, A. E. (1981). Beam induced decomposition and sputtering of LiI, *Radiat. Eff. Defects Solids* **54**, p. 135.

Szymonski, M., Kolodziej, J., Czuba, P., Piatkowski, P., Poradzist, A., Tolk, N. and Fine, J. (1991). New mechanismfor electron-stimulated desorption of nonthermal halogen atoms from alkali-halide surfaces, *Phys. Rev. Lett.* **67**, p. 1906.

Szymonski, M., Kolodziej, J., Such, B., Piatkowski, P., Struski, P., Czuba, P. and Krok, F. (2001). Nano-scale modification of ionic surfaces induced by electronic transitions, *Prog. Surf. Sci.* **67**, p. 123.

Szymonski, M., Overeijnder, H. and de Vries, A. E. (1978). Sputtering processes during 6 keV xe ion-beam bombardment of halides, *Radiat. Eff. Defects Solids* **36**, p. 189.

Szymonski, M., Struski, P., Siegel, A., Kolodziej, J. J., Such, B., Piatkowski, P., Czuba, P. and Krok, F. (2002). Ionic crystal decomposition with light, *Acta Phys. Pol.* **33**, p. 2237.

Takakusagi, S., Fukui, K., Nariyuki, F. and Iwasawa, Y. (2003a). STM study on structures of two kinds of wide strands formed on $TiO_2(110)$, *Surf. Sci.* **523**, p. L41.

Takakusagi, S., Fukui, K., Tero, R., Nariyuki, F. and Iwasawa, Y. (2003b). Self-limiting growth of Pt nanoparticles from MeCpPtMe3 adsorbed on $TiO_2(110)$ studied by scanning tunneling microscopy, *Phys. Rev. Lett.* **91**, p. 066102.

Tasker, P. W. (1979). The stability of ionic crystal surfaces, *J. Phys. C: Solid State Phys.* **12**, p. 4977.

Thompson, M. W. (1968). Energy spectrum of ejected atoms during high energy sputtering of gold, *Philos. Mag.* **18**, p. 377.

Thompson, M. W. and Nelson, R. S. (1962). Evidence for heated spikes in bombarded gold from energy spectrum of atoms ejected by 43 keV A^+ and Xe^+ ions, *Philos. Mag.* **7**, p. 2015.

Tomanek, D., Zhong, W. and Thomas, H. (1991). Calculation of an atomically modulated friction force in atomic-force microscopy, *Europhys. Lett.* **15**, p. 887.

Tomlinson, G. A. (1929). A molecular theory of friction, *Philos. Mag. Ser.* **7**, p. 905.

Tong, X., Benz, L., Kemper, P., Metiu, H., Bowers, M. T. and Buratto, S. K. (2005). Intact size-selected Au_n clusters on a $TiO_2(110)$-(1×1) surface at room temperature, *J. Am. Chem. Soc.* **127**, p. 13516.

Torbrügge, S., Reichling, M., Ishiyama, A., Morita, S. and Custance, O. (2007). Evidence of subsurface oxygen vacancy ordering on reduced $CeO_2(111)$, *Phys. Rev. Lett.* **99**, p. 056101.

Townsend, P. D. and Kelly, J. C. (1968). Slow electron induced defects in alkali halides, *Phys. Lett. A* **26**, p. 138.

Trevethan, T., Kantorovich, L., Polesel-Maris, J. and Gauthier, S. (2007a). A comparison of dynamic atomic force microscope set-ups for performing atomic scale manipulation experiments, *Nanotechnology* **18**, p. 084017.

Trevethan, T., Kantorovich, L., Polesel-Maris, J., Gauthier, S. and Shluger, A. (2007b). Multiscale model of the manipulation of single atoms on insulating surfaces using an atomic force microscope tip, *Phys. Rev. B* **76**, p. 085414.

Trevethan, T., Watkins, M., Kantorovich, L. N. and Shluger, A. L. (2007c). Controlled manipulation of atoms in insulating surfaces with the virtual atomic force microscope, *Phys. Rev. Lett.* **98**, p. 021801.

Valbusa, U., Boragno, C. and de Mongeot, F. (2002). Nanostructuring surfaces by ion sputtering, *J. Phys. Condens. Matter* **14**, p. 8153.

van Dijken, S., Jorritsma, L. C. and Poelsema, B. (1999). Grazing-incidence metal deposition: Pattern formation and slope selection, *Phys. Rev. Lett.* **82**, p. 4038.

Venables, J., Spiller, G. and Hanbücken, M. (1984). Nucleation and growth of thin-films, *Rep. Prog. Phys.* **47**, p. 399.

Viernow, J., Petrovykh, D., Kirakosian, A., Lin, J., Men, F., Henzler, M. and Himpsel, F. (1999a). Chemical imaging of insulators by STM, *Phys. Rev. B* **59**, p. 10356.

Viernow, J., Petrovykh, D., Men, F., Kirakosian, A., Lin, J. and Himpsel, F. (1999b). Linear arrays of CaF_2 nanostructures on Si, *Appl. Phys. Lett.* **74**, p. 2125.

Watkins, M. B., Trevethan, T., Shluger, A. L. and Kantorovich, L. N. (2007). Dynamical processes at oxide surfaces studied with the virtual atomic force microscope, *Phys. Rev. B* **76**, p. 245421.

Weissenrieder, J., Kaya, S., Lu, J. L., Gao, H. J., Shaikhutdinov, S., Freund, H. J., Sierka, M., Todorova, T. K. and Sauer, J. (2005). Atomic structure of a thin silica film on a Mo(112) substrate: A two-dimensional network of SiO$_4$ tetrahedra, *Phys. Rev. Lett.* **95**, p. 076103.

Winterbon, K. (1975). *Ion Implantation Range and Energy Deposition Distributions* (Plenum, New York).

Wurz, P., Sarnthein, J., Husinsky, W., Betz, G., Nordlander, P. and Wang, Y. (1991). Electron-stimulated desorption of neutral lithium atoms from LiF due to excitation of surface excitons, *Phys. Rev. B* **43**, p. 6729.

Wyder, U., Baratoff, A., Meyer, E., Kantorovich, L. N., David, J., Maier, S., Filleter, T. and Bennewitz, R. (2007). Interpretation of atomic friction experiments based on atomistic simulations, *J. Vac. Sci. Technol. B* **25**, p. 1547.

Xu, L., Henkleman, G., Campbell, C. T. and Jonsson, H. (2006). Pd diffusion on MgO(100): The role of defects and small cluster mobility, *Surf. Sci.* **600**, p. 1351.

Yamada, H., Fukuma, T., Umeda, K., Kobayashi, K. and Matsushige, K. (2002). Local structures and electrical properties of organic molecular films investigated by non-contact atomic force microscopy, *Appl. Surf. Sci.* **188**, p. 391.

Yamada, T. and Miura, K. (1998). Water adsorption on electron irradiated NaF(001) surface, *Appl. Surf. Sci.* **130-132**, p. 598.

Yamamoto, K., Iijima, T., Kunishi, T., Fuwa, K. and Osaka, T. (1989). The growth forms of small Au particles grown on KBr and NaCl substrates having monatomic steps, *J. Cryst. Growth* **94**, p. 629.

Yase, K., Ara-Kato, N., Hanada, T., Takiguchi, H., Yoshida, Y., Back, G., Abe, K. and Tanigaki, N. (1998). Aggregation mechanism in fullerene thin films on several substrates, *Thin Solid Films* **331**, p. 131.

Yu, M., Grishkowsky, D. and Balant, A. (1981). Measurement of the velocity distribution of sputtered Na atoms from NaCl by Doppler-shift laser fluorescence, *Appl. Phys. Lett.* **39**, p. 703.

Zema, N., Piacentini, M., Czuba, P., Kolodziej, J., Piatkowski, P., Postawa, Z. and Szymonski, M. (1997). Spectroscopic behavior of halogen photodesorption from alkali halides under UV and VUV excitation, *Phys. Rev. B* **55**, p. 5448.

Zimmerli, L., Maier, S., Glatzel, T., Gnecco, E., Pfeiffer, O., Diederich, F., Fendt, L., and Meyer, E. (2005). Formation of molecular wires on nanostructured KBr, *J. Phys. Confer. Series* **19**, p. 166.

Zykova-Timan, T., Ceresoli, D. and Tosatti, E. (2007). Peak effect versus skating in high-temperature nanofriction, *Nat. Mater.* **6**, p. 230.

Index